SOLIDWORKS 2020
Quick Start

David C. Planchard
CSWP & SOLIDWORKS Accredited Educator

SDC
PUBLICATIONS

SDC Publications
P.O. Box 1334
Mission, KS 66222
913-262-2664
www.SDCpublications.com
Publisher: Stephen Schroff

Download all needed model files (SOLIDWORKS 2020 folder) from the SDC Publication website (www.SDCpublications.com/downloads/978-1-63057-323-2).

ISBN-13: 978-1-63057-323-2
ISBN-10: 1-63057-323-X

Printed and bound in the United States of America.

INTRODUCTION[i]

SOLIDWORKS® 2020 Quick Start is a great starting point for those new to SOLIDWORKS or as a teaching aid in classroom training to become familiar with the software interface, basic commands and strategies as users complete a series of models while learning different ways to accomplish a particular task.

Desired outcomes and usage competencies are listed for each chapter. The book is divided into three sections with 7 chapters.

Chapter 1 - 5: Explore the SOLIDWORKS User Interface and CommandManager, Document and System properties, simple and complex parts and assemblies, proper design intent, configurations, multi-view drawings, and BOMs using basic and advanced features.

Develop a mini Stirling Engine and investigate the proper design intent and constraints. Investigate view orientation based on the Front, Top and Right planes. As you sketch, use tools such as the circle, line, centerline, slot and mirror. Modify sketches and sketch planes. Learn to set document properties, identify sketch states and insert geometric relations and dimensions along with applying proper design intent and technique.

Dynamics

Stress Analysis

Machine Design

Thermodynamics

Fluid Mechanics

Kinematics

Heat Transfer

Statics

Parts are made up of features. With features you add or remove material. You apply and edit the Extruded Boss/Base, Revolved Boss/Base, Extruded Cut, Circular Pattern and Fillet feature. You explore the Hole Wizard feature with manufacturing parameters and the Mass Properties, Measure, and Appearance tool.

Assemblies are made up of components and sub-assemblies. Incorporate a series of provided parts and your own part to create two assemblies utilizing the Bottom-up approach.

Utilize the following assembly tools: Insert Component, Mate, Hide, Show, Rotate, Move, Modify, Flexible, Ridge, Multiple mate and more. Learn how to add constraints that result in dynamic behavior of the assembly such as linear translation and rotation.

Before you machine or create a rapid prototype of a part or assembly, you should verify the model for clearance, interference, static and dynamic behavior. Verify behavior between the Power Piston, Power Clevis, Connecting Rod and Handle with design modifications using the Stirling Engine Modified assembly.

Apply the Assembly Visualization tool to sort components by mass while creating a motion study (animation) for a formal presentation.

Create two new drawings: Fly Wheel assembly and Bushing part. The first drawing is an Isometric Exploded view of the Fly Wheel assembly. The assembly drawing displays a Bill of Materials at the part level along with Balloons, Magnetic lines, and Custom Properties.

The second drawing is a Bushing part. Insert standard Orthographic (Front, Top & Right) views along with an Isometric view. Insert all needed dimensions, annotations and Custom Properties.

Chapter 6: Review the Certified SOLIDWORKS Associate (CSWA) exam. Understand the curriculum and categories of the CSWA exam and the required model knowledge needed to successfully take and pass the exam.

Exam categories are: Drafting Competencies, Basic Part Creation and Modification, Intermediate Part Creation and Modification, Advanced Part Creation and Modification, and Assembly Creation and Modification.

Chapter 7: Provide a basic understanding between Additive vs. Subtractive manufacturing. Discuss Fused Filament Fabrication (FFF), STereoLithography (SLA), and Selective Laser Sintering (SLS) printer technology. Select suitable filament material. Comprehend 3D printer terminology. Knowledge of preparing, saving, and printing a model on a Fused Filament Fabrication 3D printer. Information on the Certified SOLIDWORKS Additive Manufacturing Certification (CSWA-AM) exam.

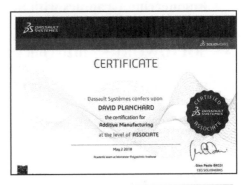

Download all needed model files (SOLIDWORKS 2020 folder) from the SDC Publications website (www.SDCpublications.com/downloads/978-1-63057-323-2) to a local hard drive.

About the Author

David Planchard is the founder of D&M Education LLC. Before starting D&M Education, he spent over 35 years in industry and academia holding various engineering, marketing, and teaching positions. He holds five U.S. patents. He has published and authored numerous papers on Machine Design, Product Design, Mechanics of Materials, and Solid Modeling. He is an active member of the SOLIDWORKS Users Group and the American Society of Engineering Education (ASEE). David holds a BSME, MSM with the following professional certifications: CCAI, CCNP, CSWA-SD, CSWSA-FEA, CSWA-AM, CSWP, CSWP-DRWT and SOLIDWORKS Accredited Educator. David is a SOLIDWORKS Solution Partner, an Adjunct Faculty member and the SAE advisor at Worcester Polytechnic Institute in the Mechanical Engineering department. In 2012, David's senior Major Qualifying Project team (senior capstone) won first place in the Mechanical Engineering department at WPI. In 2014, 2015 and 2016, David's senior Major Qualifying Project teams won the Provost award in the Mechanical Engineering department for design excellence. In 2018, David's senior Major Qualifying Project team (Co-advisor) won the Provost award in the

Electrical and Computer Engineering department. Subject area: Electrical System Implementation of Formula SAE Racing Platform.
David Planchard is the author of the following books:

- **SOLIDWORKS® 2020 Reference Guide**, 2019, 2018, 2017, 2016, 2015, 2014, 2013, 2012, 2011, 2010, and 2009

- **Engineering Design with SOLIDWORKS® 2020**, 2019, 2018, 2017, 2016, 2015, 2014, 2013, 2012, 2011, 2010, 2009, 2008, 2007, 2006, 2005, 2004, and 2003

- **Engineering Graphics with SOLIDWORKS® 2020**, 2019, 2018, 2017, 2016, 2015, 2014, 2013, 2012, and 2011

- **SOLIDWORKS® 2020 Quick Start**, 2019, 2018

- **SOLIDWORKS® 2017 in 5 Hours with video instruction**, 2016, 2015, and 2014

- **SOLIDWORKS® 2020 Tutorial**, 2019, 2018, 2017, 2016, 2015, 2014, 2013, 2012, 2011, 2010, 2009, 2008, 2007, 2006, 2005, 2004, and 2003

- **Drawing and Detailing with SOLIDWORKS® 2014**, 2012, 2010, 2009, 2008, 2007, 2006, 2005, 2004, 2003, and 2002

- **Official Certified SOLIDWORKS® Professional (CSWP) Certification Guide with video instruction, Version 5: 2018 - 2020**, Version 4: 2015 - 2017, Version 3: 2012 - 2014, Version 2: 2012 - 2013, Version 1: 2010 - 2010

- **Official Guide to Certified SOLIDWORKS® Associate Exams: CSWA, CSWA-SD, CSWSA-FEA, CSWA-AM Version 4: 2017 - 2019**, Version 3: 2015 - 2017, Version 2: 2012 - 2015, Version 1: 2012 -2013

- **Assembly Modeling with SOLIDWORKS® 2012**, 2010, 2008, 2006, 2005-2004, 2003, and 2001Plus

- **Applications in Sheet Metal Using Pro/SHEETMETAL & Pro/ENGINEER**

Acknowledgements

Writing this book was a substantial effort that would not have been possible without the help and support of my loving family and of my professional colleagues. I would like to thank Professor John M. Sullivan Jr., Professor Jack Hall and the community of scholars at Worcester Polytechnic Institute who have enhanced my life, my knowledge and helped to shape the approach and content to this text.

The author is greatly indebted to my colleagues from Dassault Systèmes SOLIDWORKS Corporation for their help and continuous support: Mike Puckett, Avelino Rochino, Yannick Chaigneau, Terry McCabe and the SOLIDWORKS Partner team.

Thanks also to Professor Richard L. Roberts of Wentworth Institute of Technology, Professor Dennis Hance of Wright State University, Professor Jason Durfess of Eastern Washington University and Professor Aaron Schellenberg of Brigham Young University - Idaho who provided vision and invaluable suggestions.

SOLIDWORKS certification has enhanced my skills and knowledge and that of my students. Thank you to Ian Matthew Jutras (CSWE) who is a technical contributor and Stephanie Planchard, technical procedure consultant.

Contact the Author

We realize that keeping software application books current is imperative to our customers. We value the hundreds of professors, students, designers, and engineers that have provided us input to enhance our book. We value your suggestions and comments. Please contact me with any comments, questions or suggestions on this book or any of our other SOLIDWORKS books. David Planchard, D & M Education, LLC, dplanchard@msn.com.

Note to Instructors

Please contact the publisher **www.SDCpublications.com** for additional classroom support materials: PowerPoint presentations, Adobe files along with avi files, additional design projects, quizzes with initial and final SOLIDWORKS models and tips that support the usage of this text in a classroom environment.

Trademarks, Disclaimer and Copyrighted Material

SOLIDWORKS®, eDrawings®, SOLIDWORKS Simulation and SOLIDWORKS Flow Simulation are registered trademarks of Dassault Systèmes SOLIDWORKS Corporation in the United States and other countries; certain images of the models in this publication courtesy of Dassault Systèmes SOLIDWORKS Corporation.

Microsoft Windows®, Microsoft Office® and its family of products are registered trademarks of the Microsoft Corporation. Other software applications and parts described in this book are trademarks or registered trademarks of their respective owners.

The publisher and the author make no representations or warranties with respect to the accuracy or completeness of the contents of this work and specifically disclaim all warranties, including without limitation warranties of fitness for a particular purpose.

No warranty may be created or extended by sales or promotional materials. Dimensions of parts are modified for illustration purposes. Every effort is made to provide an accurate text. The author and the manufacturers shall not be held liable for any parts, components, assemblies or drawings developed or designed with this book or any responsibility for inaccuracies that appear in the book. Web and company information was valid at the time of this printing.

References

- SOLIDWORKS Users Guide, SOLIDWORKS Corporation, 2020

- ASME Y14 Engineering Drawing and Related Documentation Practices

- Beers & Johnson, Vector Mechanics for Engineers, 6th ed. McGraw Hill, Boston, MA

- Betoline, Wiebe, Miller, Fundamentals of Graphics Communication, Irwin, 1995

- Hibbler, R.C, Engineering Mechanics Statics and Dynamics, 8th ed, Prentice Hall, Saddle River, NJ

- Hoelscher, Springer, Dobrovolny, Graphics for Engineers, John Wiley, 1968

- Jensen, Cecil, Interpreting Engineering Drawings, Glencoe, 2002

- Jensen & Helsel, Engineering Drawing and Design, Glencoe, 1990

- Lockhart & Johnson, Engineering Design Communications, Addison Wesley, 1999

- Olivo C., Payne, Olivo, T, Basic Blueprint Reading and Sketching, Delmar, 1988

- Planchard & Planchard, Drawing and Detailing with SOLIDWORKS, SDC Publications, Mission, KS 2014

During the initial SOLIDWORKS installation, you are requested to select either the ISO or ANSI drafting standard. ISO is typically a European drafting standard and uses First Angle Projection. The book is written using the ANSI (US) overall drafting standard and Third Angle Projection for drawings.

TABLE OF CONTENTS

Exclusive Online Bonus Content

Instructions for download on inside front cover of book

Download all needed model files (SOLIDWORKS 2020 folder) from the SDC Publications website (www.SDCpublications.com/downloads/978-1-63057-323-2).

The book provides information on creating and storing special Part, Assembly and Drawing templates in the MY-TEMPLATES folder. The MY-TEMPLATES folder is added to the New SOLIDWORKS Document dialog box. Talk to your IT department *before you set* any new locations on a network system. The procedure in the book is designed for your personal computer.

If you do not create the MY-TEMPLATE tab or the special part, drawing, or assembly templates, use the standard SOLIDWORKS default template and apply all of the needed document and custom properties.

Overview of Chapters

Chapter 1: Overview of SOLIDWORKS and the User Interface

SOLIDWORKS is a design software application used to create 2D and 3D sketches, 3D parts and assemblies and 2D drawings.

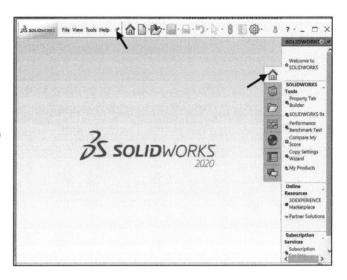

Chapter 1 introduces the user to the SOLIDWORKS User Interface (UI) and CommandManager: Menu bar toolbar, Menu bar menu, Drop-down menus, Context toolbars, Consolidated drop-down toolbars, System feedback icons, Confirmation Corner, Heads-up View toolbar, Document Properties and more.

How do you start a SOLIDWORKS session? How do you open a new or existing part? How do you start a model in SOLIDWORKS? What is design intent?

Chapter 2: 2D Sketching, Features and Parts

Learn about 2D Sketching and 3D features. Create a new part called Wheel with user defined document properties.

Create the Wheel for the Fly Wheel sub-assembly. Utilize the Fly Wheel sub-assembly in the final Stirling Engine assembly.

Apply the following sketch and feature tools: Circle, Line Centerline, Centerpoint Straight Slot, Mirror Entities, Extruded Boss, Extruded Cut, Revolved Boss, Circular Pattern, Hole Wizard and Fillet.

Incorporate design change into a part using proper design intent, along with applying multiple geometric relations: Coincident, Vertical, Horizontal, Tangent and Midpoint and feature and sketch modifications.

Utilize the Material, Mass Properties and Appearance tool on the Wheel.

Chapter 3: Assembly Modeling - Bottom-up method

Learn about the Bottom-up assembly method and create two new assemblies with user defined document properties:

- Fly Wheel.

- Stirling Engine.

Insert the following Standard and Quick mate types: Coincident, Concentric, Distance and Tangent.

Utilize the following assembly tools: Insert Component, Suppress, Un-suppress, Mate, Move Component, Rotate Component, Interference Detection, Hide, Show, Flexible, Ridge, and Multiple mate mode.

Create an Exploded View with animation.

Apply the Measure and Mass Properties tool to modify a component in the Stirling Engine assembly.

Chapter 4: Design Modifications

Address clearance, interference, static and dynamic behavior of the Stirling Engine Modified assembly.

Verify the behavior between the following components: Power Piston, Power Clevis, Connecting Rod and Handle in the assembly.

Apply the following assembly tools: Move, Rotate, Collision Detection, Interference Detection, Selected Components, Edit Feature and Center of Mass.

Utilize the Assembly Visualization tool on the Stirling Engine assembly and sort by component mass.

Create a new Coordinate System on the Stirling Engine assembly relative to the default origin.

Run a Motion Study and save the Motion Study AVI file.

Chapter 5: Drawing and Dimensioning Fundamentals

Learn about Drawing and Dimension Fundamentals and create two new drawings with user defined document properties:

- Fly Wheel Assembly.

- Bushing.

Create the Fly Wheel Assembly drawing with an Exploded Isometric view.

Utilize a Bill of Materials, Magnetic lines and Balloons.

Learn about Custom Properties and the Title Block.

Create the Bushing Part drawing utilizing Third Angle Projection with two standard Orthographic views: Front, Top and an Isometric view.

Address imported dimensions from the Model Items tool.

Insert additional dimensions using the Smart Dimension tool along with all needed annotations.

Chapter 6: Introduction to the Certified SOLIDWORKS Associate (CSWA) Exam

DS SOLIDWORKS Corp. offers various stages of certification representing increasing levels of expertise in 3D CAD design as it applies to engineering: Certified SOLIDWORKS Associate CSWA, Certified SOLIDWORKS Professional CSWP and Certified SOLIDWORKS Expert CSWE along with specialty fields in Drawing, Simulation, Sheet Metal, Surfacing and more.

The CSWA certification indicates a foundation in and apprentice knowledge of 3D CAD design and engineering practices and principles.

The CSWA Academic exam is provided either in a single 3 hour segment, or 2 - 90 minute segments.

Part 1 of the CSWA Academic exam is 90 minutes, minimum passing score is 80, with 6 questions. There are two questions in the Basic Part Creation and Modification category, two questions in the Intermediate Part Creation and Modification category and two questions in the Assembly Creation and Modification category.

Part 2 of the CSWA Academic exam is 90 minutes, minimum passing score is 80 with 8 questions. There are three questions on the CSWA Academic exam in the Drafting Competencies category, three questions in the Advanced Part Creation and Modification category and two questions in the Assembly Creation and Modification category.

The CSWA exam for industry is only provided in a single 3 hour segment. The exam consists of 14 questions and five categories.

All exams cover the same material.

To obtain additional CSWA exam information, visit the SOLIDWORKS VirtualTester Certification site at https://SOLIDWORKS.virtualtester.com/.

Chapter 7: Additive Manufacturing - 3D Printing

Provide a basic understanding between Additive vs. Subtractive manufacturing. Discuss Fused Filament Fabrication (FFF), STereoLithography (SLA), and Selective Laser Sintering (SLS) printer technology. Select suitable filament material. Comprehend 3D printer terminology. Knowledge of preparing, saving, and printing a model on a Fused Filament Fabrication 3D printer. Information on the SOLIDWORKS Additive Manufacturing Certification (CSWA-AM).

On the completion of this chapter, you will be able to:

* Discuss Additive vs Subtractive manufacturing.

* Review 3D printer technology: Fused Filament Fabrication (FFF), STereoLithography (SLA), and Selective Laser Sintering (SLS).

- Select the correct filament material:
  - o PLA (Polylactic acid)
  - o FPLA (Flexible Polylactic acid)
  - o ABS (Acrylonitrile butadiene styrene)
  - o PVA (Polyvinyl alcohol)
  - o Nylon 618
  - o Nylon 645

- Create an STL (*.stl) file, an Additive Manufacturing (*.amf) file and a 3D Manufacturing format (*.3mf) file.

- Prepare G-code.

- Comprehend general 3D printer terminology.

- Understand optimum build orientation.

- Enter slicer parameters:
  - o Raft, brim, skirt, layer height, percent infill, infill pattern, wall thickness, fan speed, print speed, bed temperature, and extruder (hot end) temperature.

- Address fit tolerance for interlocking parts.

- Define general 3D Printing tips.

- Print directly from SOLIDWORKS.

- Knowledge of the Certified SOLIDWORKS Additive Manufacturing (CSWA-AM) exam.

Book Layout

The following conventions are used throughout this book:

- The term document is used to refer to a SOLIDWORKS part, drawing or assembly file.

- The list of items across the top of the SOLIDWORKS interface is the Menu bar menu or the Menu bar toolbar. Each item in the Menu bar has a pull-down menu. When you need to select a series of commands from these menus, the following format is used: Click **View**, **Hide/Show**, check **Origins** from the Menu bar. The Origins are displayed in the Graphics window.

- The ANSI overall drafting standard and Third Angle projection is used as the default setting in this text. MMGS (millimeter, gram, second) unit system is used.

- The book is organized into various chapters. Each chapter is focused on a specific subject or feature.

- All templates, logos and needed model documents for this book are provided.

- Download all needed model files (SOLIDWORKS 2020 folder) from the SDC Publications website (www.SDCpublications.com/downloads/978-1-63057-323-2). View the models to enhance the user experience.

- Screen shots in the book were made using SOLIDWORKS 2020 SP0 running Windows® 10.

| Bracket |
| Chapter 2 Homework |
| Chapter 3 Homework |
| Chapter 4 Homework |
| Chapter 5 Homework |
| Chapter 6 CSWA Models |
| Design Modifications |
| Drawing |
| FLY WHEEL |
| LOGO |
| STIRILING ENGINE |

The following command syntax is used throughout the text. Commands that require you to perform an action are displayed in **Bold** text.

Format:	Convention:	Example:
Bold	<ul><li>All commands actions.</li><li>Selected icon button.</li><li>Selected geometry: line, circle.</li><li>Value entries.</li></ul>	<ul><li>Click **Options** ⚙️ ▾ from the Menu bar toolbar.</li><li>Click **Corner Rectangle** ▭ ▾ from the Sketch toolbar.</li><li>Click **Sketch** from the Context toolbar.</li><li>Select the **centerpoint**.</li><li>Enter **3.0** for Radius.</li></ul>
Capitalized	<ul><li>Filenames.</li><li>First letter in a feature name.</li></ul>	<ul><li>Save the **FLATBAR** assembly.</li><li>Click the **Fillet** feature.</li></ul>

Windows Terminology in SOLIDWORKS

The mouse buttons provide an integral role in executing SOLIDWORKS commands. The mouse buttons execute commands, select geometry, display Shortcut menus and provide information feedback.

A summary of mouse button terminology is displayed below:

Item:	Description:
Click	Press and release the left mouse button.
Double-click	Double press and release the left mouse button.
Click inside	Press the left mouse button. Wait a second, and then press the left mouse button inside the text box. Use this technique to modify Feature names in the FeatureManager design tree.
Drag	Point to an object, press and hold the left mouse button down. Move the mouse pointer to a new location. Release the left mouse button.
Right-click	Press and release the right mouse button. A Shortcut menu is displayed. Use the left mouse button to select a menu command.
Tool Tip	Position the mouse pointer over an Icon (button). The tool name is displayed below the mouse pointer.
Large Tool Tip	Position the mouse pointer over an Icon (button). The tool name and a description of its functionality are displayed below the mouse pointer.
Mouse pointer feedback	Position the mouse pointer over various areas of the sketch, part, assembly or drawing. The cursor provides feedback depending on the geometry.

A mouse with a center wheel provides additional functionality in SOLIDWORKS. Roll the center wheel downward to enlarge the model in the Graphics window. Hold the center wheel down. Drag the mouse in the Graphics window to rotate the model.

Visit SOLIDWORKS website: http://www.SOLIDWORKS.com /sw/support/hardware.html to view their supported operating systems and hardware requirements.

The book is designed to expose the new user to numerous tools and procedures. It may not always use the simplest and most direct process.

Hardware & System Requirements
Research graphics cards hardware, system requirements, and other related topics.

SolidWorks System Requirements
Hardware and system requirements for SolidWorks 3D CAD products.

Data Management System Requirements
Hardware and system requirements for SolidWorks Product Data Management (PDM) products.

SolidWorks Composer System Requirements
Hardware and system requirements for SolidWorks Composer and other 3DVIA related products.

SolidWorks Electrical System Requirements
Hardware and system requirements for SolidWorks Electrical products.

Graphics Card Drivers
Find graphics card drivers for your system to ensure system performance and stability.

Anti-Virus
The following Anti-Virus applications have been tested with SolidWorks 3D CAD products.

Hardware Benchmarks
Applications and references that can help determine hardware performance.

The Instructor's information contains over 60 classroom presentations, along with helpful hints, What's new, sample quizzes, avi files of assemblies, projects, and all initial and final SOLIDWORKS model files.

The book does not cover starting a SOLIDWORKS session in detail for the first time. A default SOLIDWORKS installation presents you with several options. For additional information for an Education Edition, visit the following site:
http://www.SOLIDWORKS.com/sw/engineering-education-software.htm

The CSWA certification indicates a foundation in and apprentice knowledge of 3D CAD design and engineering practices and principles.

The CSWA Academic exam is provided either in a single 3-hour segment, or 2 - 90 minute segments. The CSWA exam for industry is only provided in a single 3-hour segment. All exams cover the same material.

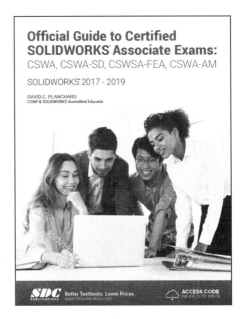

To obtain additional CSWA exam information, visit the SOLIDWORKS VirtualTester Certification site at https://SOLIDWORKS.virtualtester.com/.

.

Chapter 1

Overview of SOLIDWORKS® 2020 and the User Interface

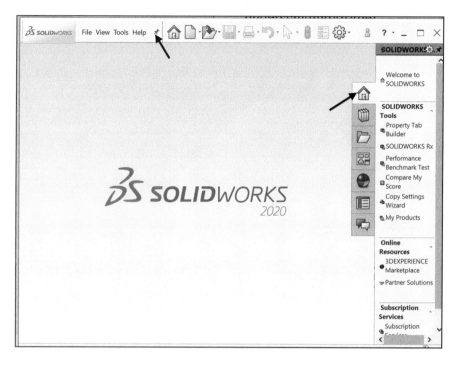

Below are the desired outcomes and usage competencies based on the completion of Chapter 1.

Desired Outcomes:	Usage Competencies:
• A comprehensive understanding of the SOLIDWORKS® 2020 User Interface (UI) and CommandManager.	• Ability to establish a SOLIDWORKS session. • Aptitude to utilize the following items: Menu bar toolbar, Menu bar menu, Drop-down menus, Context toolbars, Consolidated drop-down toolbars, System feedback icons, Confirmation Corner, Heads-up View toolbar, Document Properties, and more. • Open a new and existing SOLIDWORKS part. • Knowledge to zoom, rotate and maneuver a three-button mouse in the SOLIDWORKS Graphics window.

Notes:

Chapter 1 - Overview of SOLIDWORKS® 2020 and the User Interface

Chapter Objective

Provide a comprehensive understanding of the SOLIDWORKS® default User Interface and CommandManager: Menu bar toolbar, Menu bar menu, Drop-down menu, Right-click Pop-up menus, Context toolbars/menus, Fly-out tool button, System feedback icons, Confirmation Corner, Heads-up View toolbar, and more.

On the completion of this chapter, you will be able to:

- Utilize the SOLIDWORKS 2020 Welcome dialog box.

- Establish a SOLIDWORKS 2020 session.

- Comprehend the SOLIDWORKS 2020 User Interface.

- Recognize the default Reference Planes in the FeatureManager.

- Open a new and existing SOLIDWORKS part.

- Utilize SOLIDWORKS Help and SOLIDWORKS Tutorials.

- Zoom, rotate and maneuver a three-button mouse in the SOLIDWORKS Graphics window.

What is SOLIDWORKS®?

- SOLIDWORKS® is a mechanical design automation software package used to build parts, assemblies and drawings that takes advantage of the familiar Microsoft® Windows graphical user interface.

- SOLIDWORKS is an easy to learn design and analysis tool (SOLIDWORKS Simulation, SOLIDWORKS Motion, SOLIDWORKS Flow Simulation, Sustainability, etc.), which makes it possible for designers to quickly sketch 2D and 3D concepts, create 3D parts and assemblies and detail 2D drawings.

- Model dimensions in SOLIDWORKS are associative between parts, assemblies and drawings. Reference dimensions are one-way associative from the part to the drawing or from the part to the assembly.

Start a SOLIDWORKS 2020 Session

Start a SOLIDWORKS 2020 session and familiarize yourself with the SOLIDWORKS User Interface. As you read and perform the tasks in this chapter, you will obtain a sense of how to use the book and the structure. Actual input commands or required actions in the chapter are displayed in bold.

The book does not cover starting a SOLIDWORKS session in detail for the first time. A default SOLIDWORKS installation presents you with several options. For additional information, visit http://www.SOLIDWORKS.com.

Activity: Start a SOLIDWORKS 2020 Session.

Start a SOLIDWORKS 2020 session.

1)   Type **SOLIDWORKS 2020** in the Search window.

2)   Click the **SOLIDWORKS 2020** application (or if available, **double-click** the SOLIDWORKS icon on the desktop). The SOLIDWORKS Welcome dialog box is displayed by default.

The Welcome dialog box provides a convenient way to open recent documents (Parts, Assemblies and Drawings), view recent folders, access SOLIDWORKS resources, and stay updated on SOLIDWORKS news.

3)   **View** your options. Do not open a document at this time.

4)   If the Welcome dialog box is not displayed, click the **Welcome to SOLIDWORKS** 🏠 icon from the Standard toolbar or click **Help > Welcome to SOLIDWORKS** from the Main menu.

Home Tab

The Home tab lets you open new and existing documents, view recent documents and folders, and access SOLIDWORKS resources (*Part, Assembly, Drawing, Advanced mode, Open*).

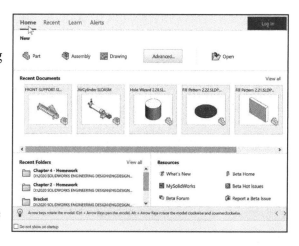

Recent Tab

The Recent tab lets you view a longer list of recent documents and folders. Sections in the Recent tab include *Documents* and *Folders*.

The Documents section includes thumbnails of documents that you have opened recently.

Click a thumbnail to open the document, or hover over a thumbnail to see the document location and access additional information about the document. When you hover over a thumbnail, the full path and last saved date of the document appears.

Learn Tab

The Learn tab lets you access instructional resources to help you learn more about the SOLIDWORKS software.

Sections in the Learn tab include:

- **Introducing SOLIDWORKS**. Open the Introducing SOLIDWORKS book.

- **Tutorials**. Open the step-by-step tutorials in the SOLIDWORKS software.

- **MySolidWorks Training**. Open the Training section at MySolidWorks.com.

- **Introducing SOLIDWORKS (Samples)**. Open local folders containing sample models.

- **Tutorials (Samples)**. Open the SOLIDWORKS Tutorials (videos) section at solidworks.com.

- **What's New (Samples)**. List of new changes.

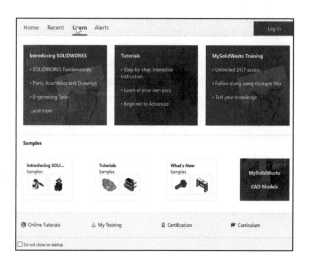

- **MySOLIDWORKS - CAD Models**. Open models in 3D ContentCentral and the Community Library.

- **My Training**. Open the My Training section at MySolidWorks.com.

- **Certification**. Open the SOLIDWORKS Certification Program section at solidworks.com.

- **Curriculum**. Open the Curriculum section at solidworks.com.

When you install the software, if you do not install the Help Files or Example Files, the Tutorials and Samples links are unavailable.

Alerts Tab

The Alerts tab keeps you updated with SOLIDWORKS news.

Sections in the Alerts tab include Critical, Troubleshooting, and Technical.

The Critical section does not appear if there are no critical alerts to display.

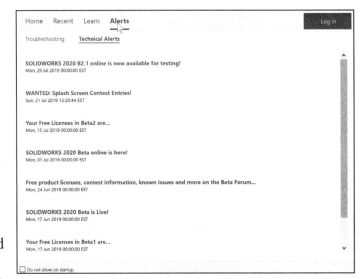

- **Troubleshooting**. Includes troubleshooting messages and recovered documents that used to be on the SOLIDWORKS Recovery tab in the Task Pane.

If the software has a technical problem and an associated troubleshooting message exists, the Welcome dialog box opens to the Troubleshooting section automatically on startup, even if you selected **Do not show at startup** in the dialog box.

- **Technical Alerts**. Open the contents of the SOLIDWORKS Support Bulletins RSS feed at solidworks.com.

Close the Welcome dialog box.

5)   Click **Close** ✕ from the Welcome dialog box. The SOLIDWORKS Graphics window is displayed. Note: You can also click outside the Welcome dialog box, in the Graphics window.

View the SOLIDWORKS Graphics window.

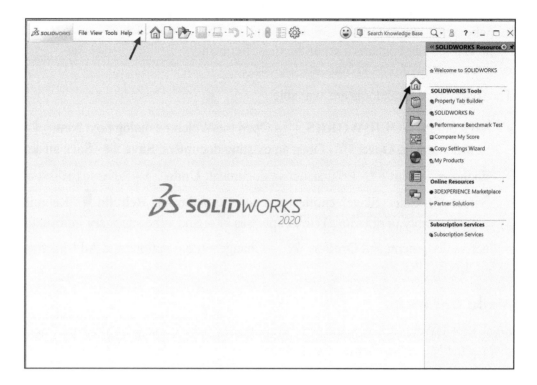

If you do not see this screen, click the SOLIDWORKS
Resources 🏠 icon on the right side of the Graphics window
located in the Task Pane.

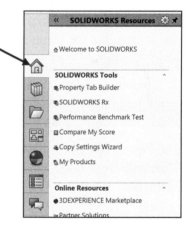

6) **Hover** the mouse pointer over the SOLIDWORKS icon.

7) **Pin** ⤚ the Menu Bar toolbar.

8) **View** your options from the Menu bar menu: **File**, **View**,
  Tools and **Help**.

Menu Bar toolbar

The SOLIDWORKS (UI) is designed to make maximum use of the Graphics window. The Menu Bar toolbar contains a set of the most frequently used tool buttons from the Standard toolbar.

The following default tools are available:

- **Welcome to SOLIDWORKS** ⌂ - Open the Welcome dialog box, **New** ▯ - Create a new document; **Open** - Open an existing document; **Save** - Save an active document; **Print** ⎙ - Print an active document; **Undo** ↺ - Reverse the last action; **Select** - Select Sketch entities, components and more; **Rebuild** - Rebuild the active part, assembly or drawing; **File Properties** - Show the summary information on the active document; and **Options** ⚙ - Change system options and Add-Ins for SOLIDWORKS.

Menu Bar menu

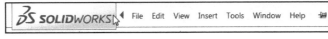

Click SOLIDWORKS in the Menu Bar toolbar to display the Menu Bar menu. SOLIDWORKS provides a context-sensitive menu structure. The menu titles remain the same for all three types of documents, but the menu items change depending on which type of document is active.

Example: The Insert menu includes features in part documents, mates in assembly documents, and drawing views in drawing documents. The display of the menu is also dependent on the workflow customization that you have selected. The default menu items for an active document are File, Edit, View, Insert, Tools, Window, Help and Pin.

The Pin ⚲ option displays the Menu bar toolbar and the Menu bar menu as illustrated. Throughout the book, the Menu bar menu and the Menu bar toolbar are referred to as the Menu bar.

Drop-down menu

SOLIDWORKS takes advantage of the familiar Microsoft® Windows user interface. Communicate with SOLIDWORKS through drop-down menus, Context sensitive toolbars, Consolidated toolbars or the CommandManager tabs.

🔅 A command is an instruction that informs SOLIDWORKS to perform a task.

To close a SOLIDWORKS drop-down menu, press the Esc key. You can also click any other part of the SOLIDWORKS Graphics window or click another drop-down menu.

Create a New Part Document

In the next section create a new part document.

Activity: Create a new Part Document.

A part is a 3D model, which consists of features. What are features?

Features are geometry building blocks.

Most features either add or remove material.

Some features do not affect material (Cosmetic Thread).

Features are created either from 2D or 3D sketched profiles or from edges and faces of existing geometry.

Features are individual shapes that combined with other features make up a part or assembly. Some features, such as bosses and cuts, originate as sketches. Other features, such as shells and fillets, modify a feature's geometry.

Features are displayed in the FeatureManager as illustrated (Boss-Extrude1, Cut-Extrude1, Cut-Extrude2, Mirror1, Cut-Extrude3 and CirPattern1).

🔅 The first sketch of a part is called the Base Sketch. The Base sketch is the foundation for the 3D model. The book focuses on 2D sketches and 3D features.

🔅 FeatureManager tabs and tree folders will vary depending on system setup and Add-ins.

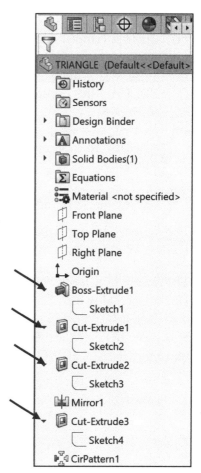

There are two modes in the New SOLIDWORKS Document dialog box: Novice and Advanced. The Novice option is the default option with three templates. The Advanced mode contains access to additional templates and tabs that you create in system options. Use the Advanced mode in this book.

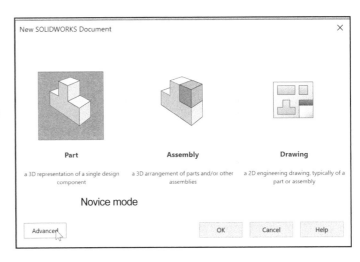

Create a new part.

9) Click **New** from the Menu bar. The New SOLIDWORKS Document dialog box is displayed.

Select the Advanced mode.

10) Click the **Advanced** button as illustrated. The Advanced mode is set.

11) Click the **Templates** tab.

12) Click **Part**. Part is the default template from the New SOLIDWORKS Document dialog box.

13) Click **OK** from the New SOLIDWORKS Document dialog box.

Illustrations may vary depending on your SOLIDWORKS version and operating system.

The Advanced mode remains selected for all new documents in the current SOLIDWORKS session. When you exit SOLIDWORKS, the Advanced mode setting is saved.

The default SOLIDWORKS installation contains three tabs in the New SOLIDWORKS Document dialog box: *Templates, MBD,* and *Tutorial*. The *Templates* tab corresponds to the default SOLIDWORKS templates. The *MBD* tab corresponds to the templates utilized in the SOLIDWORKS (Model Based Definition). The *Tutorial* tab corresponds to the templates utilized in the SOLIDWORKS Tutorials.

Part1 is displayed in the FeatureManager and is the name of the document. Part1 is the default part window name.

The Part Origin ⊥ is displayed in blue in the center of the Graphics window. The Origin represents the intersection of the three default reference planes: *Front Plane*, *Top Plane* and *Right Plane*. The positive X-axis is horizontal and points to the right of the Origin in the Front view. The positive Y-axis is vertical and points upward in the Front view. The FeatureManager contains a list of features, reference geometry, and settings utilized in the part.

💡 Edit the document units directly from the Graphics window as illustrated.

💡 CommandManager and FeatureManager tabs will vary depending on system setup and Add-ins.

View the Default Sketch Planes.

14)   Click the **Front Plane** from the FeatureManager.

15)   Click the **Top Plane** from the FeatureManager.

16)   Click the **Right Plane** from the FeatureManager.

17)   Click the **Origin** from the FeatureManager. The Origin is the intersection of the Front, Top, and Right Planes.

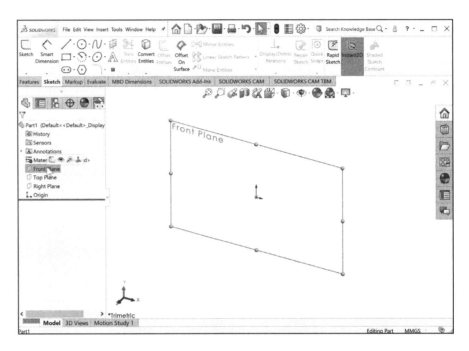

Download all needed model files (SOLIDWORKS 2020 folder) from the SDC Publication website (www.SDCpublications.com/downloads/978-1-63057-323-2). Open the Bracket part. Review the features and sketches in the Bracket FeatureManager. Work directly from a local hard drive.

Activity: Download the SOLIDWORKS folder. Open the Bracket Part.

Download the SOLIDWORKS folder. Open an existing SOLIDWORKS part.

18)   **Download** the SOLIDWORKS 2020 folder.

19)   Click **Open** from the Menu bar menu.

20)   Browse to the **SOLIDWORKS 2020\Bracket** folder.

21)   Double-click the **Bracket** part. The Bracket part is displayed in the Graphics window.

The FeatureManager design tree is located on the left side of the SOLIDWORKS Graphics window. The FeatureManager provides a summarized view of the active part, assembly, or drawing document. The tree displays the details on how the part, assembly or drawing document was created.

Use the FeatureManager rollback bar to temporarily roll back to an earlier state, to absorbed features, roll forward, roll to previous, or roll to the end of the FeatureManager design tree. You can add new features or edit existing features while the model is in the rolled-back state. You can save models with the rollback bar placed anywhere.

In the next section, review the features in the Bracket FeatureManager using the Rollback bar.

Activity: Use the FeatureManager Rollback Bar option.

Apply the FeatureManager Rollback Bar. Revert to an earlier state in the model.

22) Place the **mouse pointer** over the rollback bar in the FeatureManager design tree as illustrated. The pointer changes to a hand 🖐️. Note the provided information on the feature. This is called Dynamic Reference Visualization.

23) Drag the **rollback bar** up the FeatureManager design tree until it is above the features you want rolled back, in this case 10.0 (10) Diameter Hole1.

24) **Release** the mouse button.

View the first feature in the Bracket Part.

25) Drag the **rollback bar** up the FeatureManager above Fillet1. View the results in the Graphics window.

Return to the original Bracket Part FeatureManager.

26) Right-click **Extrude-Thin1** in the FeatureManager. The Pop-up Context toolbar is displayed.

27) Click **Roll to End**. View the results in the Graphics window.

Heads-up View toolbar

SOLIDWORKS provides the user with numerous view options. One of the most useful tools is the Heads-up View toolbar displayed in the Graphics window when a document is active.

Dynamic Annotation Views : Only available with SOLIDWORKS MBD (Model Based Definition). Provides the ability to control how annotations are displayed when you rotate models.

In the next section, apply the following tools: Zoom to Fit, Zoom to Area, Zoom out, Rotate and select various view orientations from the Heads-up View toolbar.

Activity: Utilize the Heads-up View toolbar.

Zoom to Fit the model in the Graphics window.

28) Click the **Zoom to Fit** icon. The tool fits the model to the Graphics window.

Zoom to Area on the model in the Graphics window.

29) Click the **Zoom to Area** icon. The Zoom to Area icon is displayed.

Zoom in on the top left hole.

30) **Window-select** the top left corner as illustrated. View the results.

De-select the Zoom to Area tool.

31) Click the **Zoom to Area** icon.

Fit the model to the Graphics window.

32) Press the **f** key.

Rotate the model.

33) Hold the **middle mouse button** down. Drag **upward** ↻, **downward** ↻, to the **left** ↻ and to the **right** ↻ to rotate the model in the Graphics window.

Display a few Standard Views.

34) Click **inside** the Graphics window.

35) Click **Front** from the drop-down Heads-up view toolbar. The model is displayed in the Front view.

36) Click **Right** from the drop-down Heads-up view toolbar. The model is displayed in the Right view.

37) Click **Top** from the drop-down Heads-up view toolbar. The model is displayed in the Top view.

Display a Trimetric view of the Bracket model.

38) Click **Trimetric** from the drop-down Heads-up view toolbar as illustrated. Note your options. View the results in the Graphics window.

SOLIDWORKS Help

Help in SOLIDWORKS is context-sensitive and in HTML format. Help is accessed in many ways, including Help buttons in all dialog boxes and PropertyManager (or press F1) and Help ⑦ tool on the Standard toolbar for SOLIDWORKS Help.

🔅 CommandManager and FeatureManagers tabs will vary depending on system setup and SOLIDWORKS Add-ins.

39)   Click **Help** from the Menu bar.

40)   Click **Help** ⑦ from the drop-down menu. The SOLIDWORKS Home Page is displayed by default. View your options.

💡 SOLIDWORKS Web Help is active by default under Help in the Main menu.

Close Help. Return to the SOLIDWORKS Graphics window.

41)   **Close** ☒ SOLIDWORKS Home.

SOLIDWORKS Tutorials

Display and explore the SOLIDWORKS tutorials.

42)   Click **Help** from the Menu bar.

43)   Click **Tutorials**. The SOLIDWORKS Tutorials are displayed. The SOLIDWORKS Tutorials are presented by category.

44)   Click the **Getting Started** category. The Getting Started category provides lessons on parts, assemblies, and drawings.

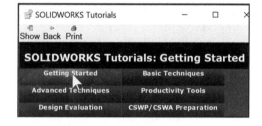

In the next section, close all models, tutorials and view the additional User Interface tools.

Activity: Close all Tutorials and Models.

Close SOLIDWORKS Tutorials and models.

45)   **Close** ☒ SOLIDWORKS Tutorials.

46)   Click **Window**, **Close All** from the Menu bar menu.

User Interface Tools

The book utilizes additional areas of the SOLIDWORKS User Interface. Explore an overview of these tools in the next section.

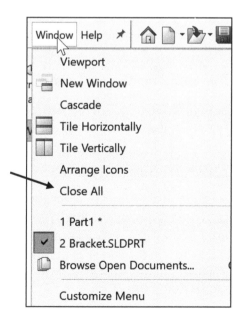

Right-click

Right-click in the Graphics window on a model, or in the FeatureManager on a feature or sketch to display the Context-sensitive toolbar. If you are in the middle of a command, this toolbar displays a list of options specifically related to that command.

Right-click an empty space in the Graphics window of a part or assembly, and a selection context toolbar above the shortcut menu is displayed. This provides easy access to the most commonly used selection tools.

Consolidated toolbar

Similar commands are grouped together in the CommandManager. For example, variations of the Rectangle sketch tool are grouped in a single fly-out button as illustrated.

If you select the Consolidated toolbar button without expanding:

For some commands such as Sketch, the most commonly used command is performed. This command is the first listed and the command shown on the button.

For commands such as rectangle, where you may want to repeatedly create the same variant of the rectangle, the last used command is performed. This is the highlighted command when the Consolidated toolbar is expanded.

System feedback icons

SOLIDWORKS provides system feedback by attaching a symbol to the mouse pointer cursor.

The system feedback symbol indicates what you are selecting or what the system is expecting you to select.

As you move the mouse pointer across your model, system feedback is displayed in the form of a symbol, riding next to the cursor as illustrated. This is a valuable feature in SOLIDWORKS.

Confirmation Corner

When numerous SOLIDWORKS commands are active, a symbol or a set of symbols is displayed in the upper right-hand corner of the Graphics window. This area is called the Confirmation Corner.

When a sketch is active, the confirmation corner box displays two symbols. The first symbol is the sketch tool icon. The second symbol is a large red X. These two symbols supply a visual reminder that you are in an active sketch. Click the sketch symbol icon to exit the sketch and to save any changes that you made.

When other commands are active, the confirmation corner box provides a green check mark and a large red X. Use the green check mark to execute the current command. Use the large red X to cancel the command.

Confirm changes you make in sketches and tools by using the D keyboard shortcut to move the OK and Cancel buttons to the pointer location in the Graphics window.

Heads-up View toolbar

SOLIDWORKS provides the user with numerous view options from the Standard Views, View and Heads-up View toolbar.

The Heads-up View toolbar is a transparent toolbar that is displayed in the Graphics window when a document is active.

For a drawing document

You can hide, move or modify the Heads-up View toolbar. To modify the Heads-up View toolbar, right-click on a tool and select or deselect the tools that you want to display.

The following views are available. Note: available views are document dependent.

- *Zoom to Fit* : Fit the model to the Graphics window.

- *Zoom to Area* : Zoom to the areas you select with a bounding box.

- *Previous View* : Display the previous view.

- *Section View* : Display a cutaway of a part or assembly, using one or more cross section planes.

- *Dynamic Annotation Views* : Only available with SOLIDWORKS MBD. Control how annotations are displayed when you rotate a model.

The Orientation dialog has an option to display a view cube (in-context View Selector) with a live model preview. This helps the user to understand how each standard view orientates the model. With the view cube, you can access additional standard views. The views are easy to understand, and they can be accessed simply by selecting a face on the cube.

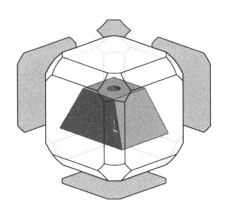

To activate the Orientation dialog box, press (Ctrl + spacebar) or click the View Orientation icon from the Heads up View toolbar. The active model is displayed in the View Selector in an Isometric orientation (default view).

Click the View Selector icon in the Orientation dialog box to show or hide the in-context View Selector.

Press **Ctrl + spacebar** to activate the View Selector.

Press the **spacebar** to activate the Orientation dialog box.

- *View Orientation box* : Select a view orientation or the number of viewports. The options are: *Top, Left, Front, Right, Back, Bottom, Single view, Two view - Horizontal, Two view - Vertical, Four view.* Click the drop-down arrow to access Axonometric views: *Isometric, Dimetric* and *Trimetric.*

- *Display Style* : Display the style for the active view. The options are: *Wireframe, Hidden Lines Visible, Hidden Lines Removed, Shaded, Shaded With Edges.*

- *Hide/Show Items* : Select items to hide or show in the Graphics window. The available items are document dependent. Note the View Center of Mass ✛ icon.

- *Edit Appearance* ⬤ : Edit the appearance of entities of the model.

- *Apply Scene* 🌐 : Apply a scene to an active part or assembly document. View the available options.

- *View Setting* 🖥 : Select the following settings: *RealView Graphics, Shadows In Shaded Mode, Ambient Occlusion, Perspective* and *Cartoon.*

- *Rotate view* ↻ : Rotate a drawing view. Input Drawing view angle and select the ability to update and rotate center marks with view.

- *3D Drawing View* 🔲 : Dynamically manipulate the drawing view in 3D to make a selection.

To display a grid for a part, click Options ⚙ ⁻, Document Properties tab. Click Grid/Snaps, check the Display grid box.

🔆 Add a custom view to the Heads-up View toolbar. Press the space key. The Orientation dialog box is displayed. Click the New View 🔍 tool. The Name View dialog box is displayed. Enter a new named view. Click OK.

Use commands to display information about the triad or to change the position and orientation of the triad. Available commands depend on the triad's context.

🔆 Save space in the CommandManager; limit your CommandManager tabs. Right-click on a CommandManager tab. Click Tabs. View your options to display CommandManager tabs.

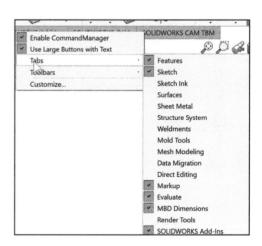

SOLIDWORKS CommandManager

The SOLIDWORKS CommandManager is a Context-sensitive toolbar. By default, it has toolbars embedded in it based on your active document type. When you click a tab below the CommandManager, it updates to display that toolbar. For example, if you click the Sketch tab, the Sketch toolbar is displayed.

☀ For commercial users, SOLIDWORKS Model Based Definition (MBD) and SOLIDWORKS CAM is a separate application. For education users, SOLIDWORKS MBD and SOLIDWORKS CAM is included in the SOLIDWORKS Education Edition.

Below is an illustrated CommandManager for a **Part** document. Tabs will vary depending on system setup and Add-ins.

☀ Set button size from the Toolbars tab of the Customize dialog box. To facilitate element selection on touch interfaces such as tablets, you can set up the larger Size buttons and text from the Options menu (Standard toolbar).

The SOLIDWORKS CommandManager is a Context-sensitive toolbar that automatically updates based on the toolbar you want to access. By default, it has toolbars embedded in it based on your active document type. The available tools are feature and document dependent.

Below is an illustrated CommandManager for a **Drawing** document. Tabs will vary depending on system setup and Add-ins.

 To add a custom tab, right-click on a tab and click Customize CommandManager. You can also select to add a blank tab and populate it with custom tools from the Customize dialog box.

The SOLIDWORKS CommandManager is a Context-sensitive toolbar that automatically updates based on the toolbar you want to access. By default, it has toolbars embedded in it based on your active document type. The available tools are feature and document dependent.

Below is an illustrated CommandManager for an **Assembly** document. Tabs will vary depending on system setup and Add-ins.

The Ink Markup toolbar is renamed to the Markup toolbar. The Markup tab is displayed by default in the CommandManager on some systems. You can draw markups with a mouse on non-touch devices, display bounding boxes for markups, create markups in drawings, and use the context toolbar to access markup options.

The Markup toolbar displays different options depending on the device. Draw and Touch are **not** available for non-touch devices.

Float the CommandManager. Drag the Features, Sketch or any CommandManager tab. Drag the CommandManager anywhere on or outside the SOLIDWORKS window.

To dock the CommandManager, perform one of the following:

While dragging the CommandManager in the SOLIDWORKS window, move the pointer over a docking icon -

Dock above , Dock left , Dock right and click the needed command.

Double-click the floating CommandManager to revert the CommandManager to the last docking position.

Screen shots in the book were made using SOLIDWORKS 2020 SP0 running Windows® 10.

Selection Enhancements

Right-click an empty space in the Graphics window of a part or assembly; a selection context toolbar above the shortcut menu provides easy access to the most commonly used selection tools.

- **Box Selection** ⬚. Select entities in parts, assemblies, and drawings by dragging a selection box with the pointer.

- **Lasso Selection** ⌇. Select entities by drawing a lasso around the entities.

- **Selection Filters** ⫶. List of selection filter commands.

- **Select Other** ⬚. Display the Select Other dialog box.

- **Select** ⬚. Display a list of selection commands.

- **Magnified Selection** ⬚. Display the magnifying glass, which gives you a magnified view of a section of a model.

Save space in the CommandManager, right-click in the CommandManager and un-check the Use Large Buttons with Text box. This eliminates the text associated with the tool.

FeatureManager Design Tree

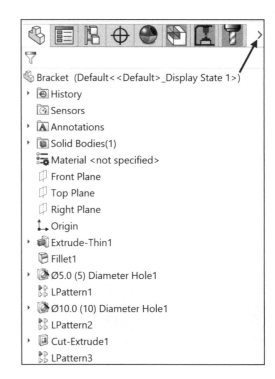

The FeatureManager consists of various tabs:

- *FeatureManager design tree* tab.

- *PropertyManager* tab.

- *ConfigurationManager* tab.

- *DimXpertManager* tab.

- *DisplayManager* tab.

- *CAM FeatureManager tree* tab.

- *CAM Operation tree* tab.

- *CAM Tools tree* tab.

Click the direction arrows to expand or collapse the FeatureManager design tree.

CommandManager and FeatureManager tabs and folder files will vary depending on system setup and Add-ins.

Select the Hide/Show FeatureManager Area tab

as illustrated to enlarge the Graphics window for modeling.

The Sensors tool located in the FeatureManager monitors selected properties in a part or assembly and alerts you when values deviate from the specified limits. There are five sensor types: Simulation Data, Mass properties, Dimensions, Measurement and Costing Data.

Various commands provide the ability to control what is displayed in the FeatureManager design tree.

1. Show or Hide FeatureManager items.

Click **Options** ⚙ from the Menu bar. Click **FeatureManager** from the System Options tab. Customize your FeatureManager from the Hide/Show tree Items dialog box.

2. Filter the FeatureManager design tree. Enter information in the filter field. You can filter by *Type of features, Feature names, Sketches, Folders, Mates, User-defined tags* and *Custom properties.*

Tags are keywords you can add to a SOLIDWORKS document to make them easier to filter and to search. The Tags 🏷 icon is located in the bottom right corner of the Graphics window.

Collapse all items in the FeatureManager, **right-click** and select **Collapse items**, or press the **Shift + C** keys.

The FeatureManager design tree and the Graphics window are dynamically linked. Select sketches, features, drawing views, and construction geometry in either pane.

Split the FeatureManager design tree and either display two FeatureManager instances, or combine the FeatureManager design tree with the ConfigurationManager or PropertyManager.

Move between the FeatureManager design tree, PropertyManager, ConfigurationManager, DimXpertManager, DisplayManager and others by selecting the tab at the top of the menu.

The ConfigurationManager tab is located to the right of the PropertyManager tab. Use the ConfigurationManager to create, select and view multiple configurations of parts and assemblies.

The icons in the ConfigurationManager denote whether the configuration was created manually or with a design table.

The DimXpertManager ⊕ tab provides the ability to insert dimensions and tolerances manually or automatically. The options are: **Auto Dimension Scheme** ⊕, **Auto Pair Tolerance** ⊞, **Basic, Location Dimension** ⊢⊣, **Basic Size Dimension** ⬒, **General Profile Tolerance** ⬓, **Show Tolerance Status** ±⊙, **Copy Scheme** ⊕, **Import Scheme** ⊕, **TolAnalyst Study** ⊠ and **Datum Target** ⊠.

💡 TolAnalyst is available in SOLIDWORKS Premium.

Fly-out FeatureManager

The fly-out FeatureManager design tree provides the ability to view and select items in the PropertyManager and the FeatureManager design tree at the same time.

Throughout the book, you will select commands and command options from the drop-down menu, fly-out FeatureManager, Context toolbar, or from a SOLIDWORKS toolbar.

💡 Another method for accessing a command is to use the accelerator key. Accelerator keys are special keystrokes, which activate the drop-down menu options. Some commands in the menu bar and items in the drop-down menus have an underlined character.

Task Pane

The Task Pane is displayed when a SOLIDWORKS session starts. You can show, hide, and reorder tabs in the Task Pane. You can also set a tab as the default so it appears when you open the Task Pane, pin or unpin to the default location.

The Task Pane contains the following default tabs:

- *SOLIDWORKS Resources* 🏠.

- *Design Library* 📦 .

- *File Explorer* 📁 .

- *View Palette* 🔲 .

- *Appearances, Scenes and Decals* ⚫ .

- *Custom Properties* 📋 .

- *SOLIDWORKS Forum* 💬 .

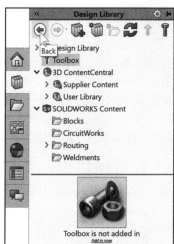

💡 Additional tabs are displayed with Add-Ins.

Use the **Back** and **Forward** buttons in the Design Library tab and the Appearances, Scenes, and Decals tab of the Task Pane to navigate in folders.

SOLIDWORKS Resources

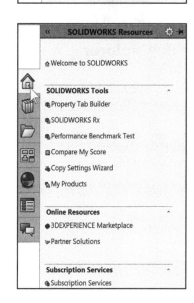

The basic SOLIDWORKS Resources 🏠 menu displays the following default selections:

- *Welcome to SOLIDWORKS.*

- *SOLIDWORKS Tools.*

- *Online Resources.*

- *Subscription Services.*

Other user interfaces are available during the initial software installation selection: *Machine Design, Mold Design, Consumer Products Design, etc.*

Design Library

The Design Library ⬚ contains reusable parts, assemblies, and other elements including library features.

The Design Library tab contains four default selections. Each default selection contains additional sub-categories.

The default selections are:

- *Design Library.*

- *Toolbox.*

- *3D ContentCentral (Internet access required).*

- *SOLIDWORKS Content (Internet access required).*

☼ Activate the SOLIDWORKS Toolbox. Click Tools, Add-Ins.., from the Main menu. Check the SOLIDWORKS Toolbox Library and SOLIDWORKS Toolbox Utilities box from the Add-ins dialog box or click SOLIDWORKS Toolbox from the SOLIDWORKS Add-Ins tab.

To access the Design Library folders in a non-network environment, click Add File Location ⬚ and browse to the needed path. Paths may vary depending on your SOLIDWORKS version and window setup. In a network environment, contact your IT department for system details.

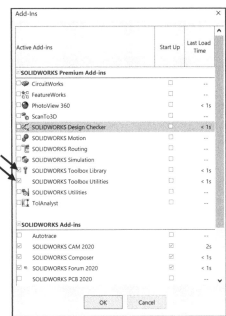

File Explorer

File Explorer ⬠ duplicates Windows Explorer from your local computer and displays:

- *Recent Documents.*

- *Samples.*

- *Open in SOLIDWORKS*

- *Desktop.*

Search

The SOLIDWORKS Search box is displayed in the upper right corner of the SOLIDWORKS Graphics window (Menu Bar toolbar). Enter the text or key words to search.

New search modes have been added to SOLIDWORKS Search as illustrated.

View Palette

The View Palette ⊞ tool located in the Task Pane provides the ability to insert drawing views of an active document, or click the Browse button to locate the desired document.

Click and drag the view from the View Palette into an active drawing sheet to create a drawing view.

🔅 The selected model is LINKAGE in the illustration.

Appearances, Scenes, and Decals

Appearances, Scenes, and Decals provide a simplified way to display models in a photo-realistic setting using a library of Appearances, Scenes, and Decals.

An appearance defines the visual properties of a model, including color and texture. Appearances do not affect physical properties, which are defined by materials.

Scenes provide a visual backdrop behind a model. In SOLIDWORKS they provide reflections on the model. PhotoView 360 is an Add-in. Drag and drop a selected appearance, scene or decal on a feature, surface, part or assembly.

Custom Properties

The Custom Properties tool provides the ability to enter custom and configuration specific properties directly into SOLIDWORKS files.

SOLIDWORKS Forum

Click the SOLIDWORKS Forum icon to search directly within the Task Pane. An internet connection is required. You are required to register and to log in for postings and discussions.

User Interface for Scaling High Resolution Screens

The SOLIDWORKS software supports high-resolution, high-pixel density displays. All aspects of the user interface respond to the Microsoft Windows® display scaling setting. In dialog boxes, PropertyManagers, and the FeatureManager design tree, the SOLIDWORKS software uses your display scaling setting to display buttons and icons at an appropriate size. Icons that are associated with text are scaled to a size appropriate for the text. In addition, for toolbars, you can display Small, Medium, or Large buttons. Click the **Options drop-down arrow** from the Standard Menu bar, and click Button size to size the icons.

Motion Study tab

Motion Studies are graphical simulations of motion for an assembly. Access the MotionManager from the Motion Study tab. The Motion Study tab is located in the bottom left corner of the Graphics window.

Incorporate visual properties such as lighting and camera perspective. Click the Motion Study tab to view the MotionManager. Click the Model tab to return to the FeatureManager design tree.

The MotionManager displays a timeline-based interface and provides the following selections from the drop-down menu as illustrated:

- *Animation:* Apply Animation to animate the motion of an assembly. Add a motor and insert positions of assembly components at various times using set key points. Use the Animation option to create animations for motion that do **not** require accounting for mass or gravity.

- *Basic Motion:* Apply Basic Motion for approximating the effects of motors, springs, collisions and gravity on assemblies. Basic Motion takes mass into account in calculating motion. Basic Motion computation is relatively fast, so you can use this for creating presentation animations using physics-based simulations. Use the Basic Motion option to create simulations of motion that account for mass, collisions or gravity.

If the Motion Study tab is not displayed in the Graphics window, click **View ➤ Toolbars ➤ MotionManager** from the Menu bar.

3D Views tab

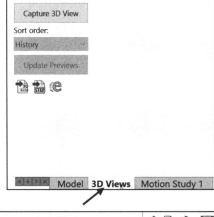

SOLIDWORKS MBD (Model Based Definition) lets you create models without the need for drawings giving you an integrated manufacturing solution. MBD helps companies define, organize, and publish 3D product and manufacturing information (PMI), including 3D model data in industry standard file formats.

Create 3D drawing views of your parts and assemblies that contain the model settings needed for review and manufacturing. This lets users navigate back to those settings as they evaluate the design.

Use the tools in the MBD Dimensions CommandManager to set up your model with selected configurations, including explodes and abbreviated views, annotations, display states, zoom level, view orientation and

section views. Capture those settings so that you and other users can return to them at any time using the 3D view palette.

To access the 3D View palette, click the 3D Views tab at the bottom of the SOLIDWORKS window or the SOLIDWORKS MBD tab in the CommandManager. The Capture 3D View button opens the Capture 3D View PropertyManager, where you specify the 3D view name, and the configuration, display state and annotation view to capture. See SOLIDWORKS help for additional information.

Dynamic Reference Visualization (Parent/Child)

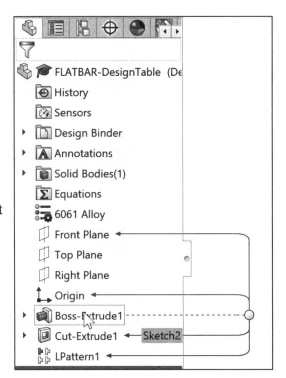

Dynamic Reference Visualization provides the ability to view the parent/child relationships between items in the FeatureManager design tree. When you hover over a feature with references in the FeatureManager design tree, arrows display showing the relationships. If a reference cannot be shown because a feature is not expanded, the arrow points to the feature that contains the reference and the actual reference appears in a text box to the right of the arrow.

Use Dynamic reference visualization for a part, assembly and mates.

To display the Dynamic Reference Visualization, click **View ➢ User Interface ➢ Dynamic Reference Visualization Parent/Child)** from the Main menu bar.

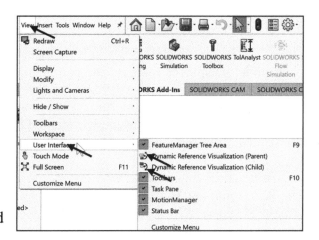

Mouse Movements

A mouse typically has two buttons: a primary button (usually the left button) and a secondary button (usually the right button). Most mice also include a scroll wheel between the buttons to help you scroll through documents and to Zoom in, Zoom out and rotate models in SOLIDWORKS. It is highly recommended that you use a mouse with at least a Primary, Scroll and Secondary button.

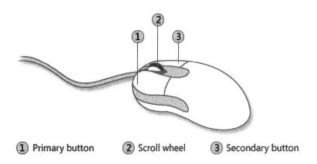

Single-click

To click an item, point to the item on the screen, and then press and release the primary button (usually the left button). Clicking is most often used to select (mark) an item or open a menu. This is sometimes called single-clicking or left-clicking.

Double-click

To double-click an item, point to the item on the screen, and then click twice quickly. If the two clicks are spaced too far apart, they might be interpreted as two individual clicks rather than as one double-click. Double-clicking is most often used to open items on your desktop. For example, you can start a program or open a folder by double-clicking its icon on the desktop.

Right-click

To right-click an item, point to the item on the screen, and then press and release the secondary button (usually the right button). Right-clicking an item usually displays a list of things you can do with the item. Right-click in the open Graphics window or on a command in SOLIDWORKS, and additional pop-up context is displayed.

Scroll wheel

Use the scroll wheel to zoom-in or to zoom-out of the Graphics window in SOLIDWORKS. To zoom-in, roll the wheel backward (toward you). To zoom-out, roll the wheel forward (away from you).

Summary

The SOLIDWORKS (UI) is designed to make maximum use of the Graphics window for your model. Displayed toolbars and commands are kept to a minimum.

The SOLIDWORKS User Interface and CommandManager consist of the following main options: Menu bar toolbar, Menu bar menu, Drop-down menus, Context toolbars, Consolidated fly-out menus, System feedback icons, Confirmation Corner and Heads-up View toolbar.

The Part CommandManager controls the display of tabs: *Features, Sketch, Markup, Evaluate, MBD Dimensions* and various *SOLIDWORKS Add-Ins*.

The FeatureManager consists of various tabs:

- *FeatureManager design tree* tab.

- *PropertyManager* tab.

- *ConfigurationManager* tab.

- *DimXpertManager* tab.

- *DisplayManager* tab.

- *CAM FeatureManager tree* tab.

- *CAM Operation tree* tab.

- *CAM Tools tree* tab.

Click the 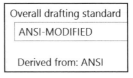 direction arrows to expand or collapse the FeatureManager design tree.

CommandManager and FeatureManager tabs and file folders will vary depending on system setup and Add-ins.

You learned about creating a new SOLIDWORKS part and opening an existing SOLIDWORKS part along with using the Rollback bar to view the sketches and features.

If you modify a document property from an Overall drafting standard, a modify message is displayed as illustrated.

Overall drafting standard
ANSI-MODIFIED
Derived from: ANSI

Use the Search box, in the upper left corner of the Materials dialog box, to search through the entire materials library.

Templates are part, drawing and assembly documents which include user-defined parameters. Open a new part, drawing or assembly. Select a template for the new document.

In Chapter 2, establish a SOLIDWORKS session. Learn about 2D Sketching and 3D features.

Create a new part. Create the Wheel for the Fly Wheel sub-assembly.

Utilize the Fly Wheel sub-assembly in the final Stirling Engine assembly.

Apply the following sketch and feature tools: Circle, Line Centerline, Centerpoint Straight Slot, Mirror Entities, Extruded Boss, Extruded Cut, Revolved Boss, Circular Pattern, Hole Wizard and Fillet.

Incorporate design change into a part using proper design intent, along with applying multiple geometric relations: Coincident, Vertical, Horizontal, Tangent and Midpoint and feature and sketch modifications.

Utilize the Material, Mass Properties and Appearance tool on the Wheel.

Notes:

Chapter 2

2D Sketching, Features and Parts

Below are the desired outcomes and usage competencies based on the completion of Chapter 2.

Desired Outcomes:	Usage Competencies:

• Create a SOLIDWORKS Part document:
 	
  o Set drafting standard.
 	
  o Set units and precision.
 	
• Apply proper design intent.
 	
• Create a part:
  o Wheel.
 	
• Utilize the Wheel in the Fly Wheel sub-assembly.
 	
• Utilize the Fly Wheel in the final Stirling Engine assembly.
	• Aptitude to establish a SOLIDWORKS session and set part document properties.
	
	• Knowledge of the following sketch and feature tools: Circle, Line, Centerline, Center point Straight Slot, Mirror Entities, Extruded Base, Revolved Boss, Extruded Cut, Circular Pattern, Hole Wizard, and Fillet.
	
	• Ability to incorporate design change into the design.
	
	• Skill to insert multiple geometric relations.
	
	• Comprehend applying material and using the Mass Properties tool.

Notes:

Chapter 2 - 2D Sketching, Features and Parts

Chapter Objective

Establish a SOLIDWORKS session. Create a new part called Wheel with user defined document properties. The Wheel part is used in the Fly Wheel sub-assembly. The sub-assembly is used in the final Stirling Engine assembly.

On the completion of this chapter, you will be able to:

- Establish a SOLIDWORKS 2020 session.

- Set user defined document properties.

- Recognize the default Sketch Planes in the Part FeatureManager.

- Select the correct Sketch plane and orientation for the Wheel.

- Utilize the following Sketch tools: Circle, Line, Centerline, Center point Straight Slot, and Mirror Entities.

- Apply and comprehend proper design intent.

- Edit a Sketch and Sketch Plane.

- Insert the following Geometric relations: Coincident, Vertical, Horizontal, Tangent, and Midpoint.

- Create and modify the following features: Extruded Base, Revolved Boss, Extruded Cut, Circular Pattern, Hole Wizard, and Fillet.

- Apply Material and utilize the Mass Properties tool to verify a design change.

- Utilize and apply the Appearance tool.

Activity: Start a SOLIDWORKS Session. Create a New Part Document.

Start a SOLIDWORKS session.

1) Double-click the **SOLIDWORKS 2020 icon** from the Desktop. **Close** the SOLIDWORKS Welcome dialog box. The SOLIDWORKS program is displayed.

2) Pin ⚲ the Menu bar menu as illustrated. Use the Menu bar and the Menu bar toolbar.

Create a New Part Document.

3) Click **New** ⬜ from the Menu bar or click **File**, **New** from the Menu bar menu. The New SOLIDWORKS Document dialog box is displayed. Advanced mode is used in this book.

4) Click the **Advanced** tab.

5) Click the **Templates** tab.

6) Double-click **Part** from the New SOLIDWORKS Document dialog box. Part 1 is displayed in the FeatureManager.

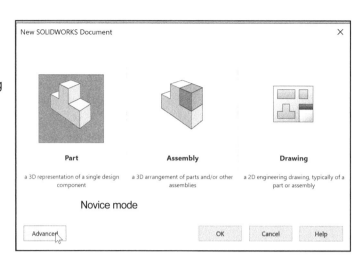

☼ The first system default Part filename is Part1. The system attaches the .sldprt suffix to the created part. The second created part in the same session increments to the filename Part2.

Display the origin.

7) Click **View**, **Hide/Show**, **Origins** from the Main menu. The origin is displayed in the Graphics window.

Directional input refers by default to the global coordinate system (X-, Y- and Z), which is based on Plane1 with its origin located at the origin of the part or assembly. Plane1 (Front) is the first plane that appears in the FeatureManager design tree and can have a different name. The reference triad shows the global X-, Y- and Z-directions.

The three default ⊥ reference planes, displayed in the FeatureManager design tree, represent infinite 2D planes in 3D space.

Activity: Set Document Properties

Set Document Properties. Set overall drafting standard.

8) Click **Options** ⚙ from the Menu bar toolbar. The System Options General dialog box is displayed.

9) Click the **Document Properties** tab.

10) Select **ANSI** from the Overall drafting standard drop-down menu.

🔆 Various detailing options are available depending on the selected standard.

The Overall drafting standard determines the display of dimension text, arrows, symbols, and spacing. Units are the measurement of physical quantities. Millimeter dimensioning and decimal inch dimensioning are the two most common unit types specified for engineering parts and drawings.

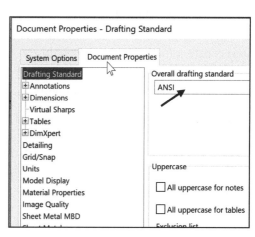

Set Document Properties. Set units and precision.

11) Click the **Units** folder.

12) Click **MMGS** (millimeter, gram, second) for Unit system.

13) Select **.12** (two decimal places) for Length basic units.

14) Click **OK** from the Document Properties - Units dialog box. The Part FeatureManager is displayed.

🔆 The origin ⌙ represents the intersection of the Front, Top and Right planes.

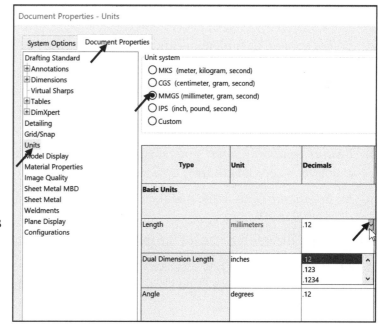

2D Sketching - Identify the Correct Sketch Plane

Most SOLIDWORKS features start with a 2D sketch. Sketches are the foundation for creating features. SOLIDWORKS provides the ability to create either 2D or 3D sketches.

A 2D sketch is limited to a flat 2D sketch plane located on a reference plane, face or a created plane. 3D sketches are very useful in creating sketch geometry that does not lie on an existing or easily defined plane.

Does it matter what plane you start the base 2D sketch on? Yes. When you create a new part or assembly, the three default planes are aligned with specific views. The plane you select for your first sketch determines the orientation of the part. Selecting the correct plane to start your model is very important.

Sketch States

Sketches can exist in any of five states. The state of the sketch is displayed in the status bar at the bottom of the SOLIDWORKS window. The following are the five sketch states in SOLIDWORKS:

1. *Under Defined*. Inadequate definition of the sketch (blue). The FeatureManager displays a minus (-) symbol before the sketch name.

2. *Fully Defined*. Complete information (black). The FeatureManager displays no symbol before the sketch name.

3. *Over Defined*. Duplicate dimensions and or relations (orange - red). The FeatureManager displays a (+) symbol before the sketch name. The What's Wrong dialog box is displayed.

4. *Invalid Solution Found*. Your sketch is solved but results in invalid geometry, for example, a zero length line, zero radius arc or a self-intersecting spline (yellow).

5. *No Solution Found*. Indicates sketch geometry that cannot be resolved (Brown).

 Color indicates the state of the individual Sketch entities.

In SOLIDWORKS, it is not necessary to fully dimension or define sketches before you use them to create features. You should fully define sketches before you consider the part finished for manufacturing.

Activity: Create the Base Sketch for the First Feature of the Wheel

In SOLIDWORKS, a 2D profile is called a sketch. The Base sketch is the first sketch in the feature. A sketch requires a sketch plane and a 2D profile. The sketch in this example uses the Front Plane. The 2D profile is a circle. A Geometric relationship and a dimension define the exact size and location of the center point of the circle relative to the origin. The center of the circle is Coincident with the origin. The origin is displayed in red.

Create the Base Sketch.

15) Right-click **Front Plane** from the FeatureManager. The Context toolbar is displayed. This is your Sketch plane. The Sketch toolbar is displayed. Front Plane is your Sketch plane.

16) Click **Sketch** ⌐ from the Context toolbar as illustrated.

💡 CommandManager and FeatureManager tabs and folder files will vary depending on system setup and Add-ins.

💡 If needed, click View, Hide/Show from the Main menu bar. Click Sketch Relations to display relations in the Graphics window.

The Front Plane rotates normal to the Sketch Plane.

17) Click the **Circle** ⊙ tool from the Sketch toolbar. The Circle PropertyManager is displayed. The sketch opens in the Front view. The mouse pointer displays the Circle symbol icon. The Front Plane feedback indicates the current Sketch plane.

The Circle-based tool uses a Consolidated Circle PropertyManager. The SOLIDWORKS application defaults to the last used tool type.

18) Drag the **mouse pointer** into the Graphics window. The cursor displays the Circle symbol icon.

19) Click the **Origin** ↳ in the Graphics window. This is the first point of the circle. It is very important that you always reference the sketch to the origin. This helps to fully define the sketch. The cursor displays a Coincident relation to the origin.

20) Drag the **mouse pointer** to the right of the origin (approximately 25mm) to create the circle as illustrated. The center point of the circle is positioned at the origin.

21) Click a **position** to create the circle. The sketch is under defined and is displayed in blue.

If needed, view the sketch relations in the Graphics window.

22) Click **View**, **Hide/Show**, **Sketch Relations** from the Main Menu bar. The sketch relation (Coincident <img>) is displayed in the Graphics window.

The diameter of the circle is displayed above the mouse pointer as you drag the mouse pointer up and to the right. The diameter displays different values. Define the exact dimension with the Smart dimension tool.

Add a dimension to fully define the sketch.

23) Click the **Smart Dimension** tool from the Sketch toolbar. The cursor displays the Smart Dimension icon.

24) Click the **circumference** of the circle.

25) Click a **position** diagonally above and to the right of the circle in the Graphics window.

26) Enter **20**mm in the Modify dialog box.

27) Click the **Green Check mark** in the Modify dialog box. The diameter of the circle is 20mm. The sketch is fully defined and is displayed in black.

⌀ **20.00**

Add relations then dimensions. This keeps the user from having too many unnecessary dimensions. This also helps to show the design intent of the model.

If needed, click View, Hide/Show from the Main menu bar. Click Sketch Relations to display relations in the Graphics window.

Features

- Features are geometry building blocks.

- Features add or remove material.

- Features are created from sketched profiles or from edges and faces of existing geometry.

What is a Base feature? The Base feature for the Wheel (Extruded Base) is the first feature that is created. The Base feature is the foundation of the part. Keep the Base feature simple.

Activity: Create the First Feature of the Wheel - Extruded Base

Create the first feature of the Wheel. Extrude the sketch.

28) Click the **Features** tab from the CommandManager. The Features toolbar is displayed.

29) Click **Extruded Boss/Base** from the Features toolbar. The Boss-Extrude PropertyManager is displayed. Blind is the default End Condition in Direction 1. The extruded sketch is previewed in a Trimetric view. The preview displays the direction of the Extrude feature.

30) Select **Mid Plane** for End Condition in Direction 1. Note your End Condition options.

31) Enter **20**mm for Depth in Direction 1. Accept the default conditions.

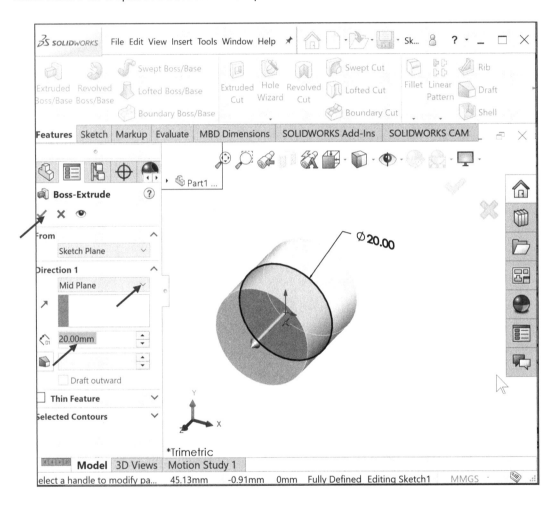

32) Click **OK** from the Boss-Extrude PropertyManager. Boss-Extrude1 is displayed in the FeatureManager.

Fit the model to the Graphics window.

33)   Press the **f** key.

Design Intent

What is design intent? All designs are created for a purpose. Design intent is the intellectual arrangements of features and dimensions of a design. Design intent governs the relationship between sketches in a feature, features in a part and parts in an assembly.

The SOLIDWORKS definition of design intent is the process in which the model is developed to accept future modifications. Models behave differently when design changes occur.

Design for change. Utilize geometry for symmetry, reuse common features, and reuse common parts. Build change into the following areas that you create: sketch, feature, part, assembly and drawing.

During the sketching process, insert all needed geometric relations and dimensions relative to the origin to fully define the sketch. Reference the origin in the sketch. The origin represents the intersection of the Front, Top and Right Planes.

Start the translation of the initial design functional and
geometric requirements into SOLIDWORKS features.

Save the part. Create a new folder. Enter name. Enter description.

34) Click **Save As** 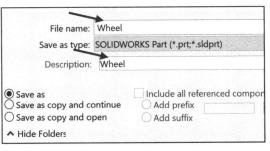 from the Drop-down
Menu bar.

35) Click the **DOCUMENTS** file folder on your
computer. Note: The procedure will be
different depending on your operating
system.

36) Click **New Folder**.

37) Enter **SOLIDWORKS 2020** for the folder
name.

38) Double-click the **SOLIDWORKS 2020**.
folder.

39) Enter **Wheel** for File name.

40) Enter **Wheel** for Description.

41) Click **Save** from the Save As dialog box. The Wheel
FeatureManager is displayed.

42) **Expand** Boss-Extrude1 from the FeatureManager as
illustrated. Sketch1 is fully defined. Sketch1 has all needed
relations and dimensions relative to the origin.

A fully defined sketch has complete information
(manufacturing and inspection) and is displayed in black.

☀ Display an Isometric view of the model. Press the **space
bar** to display the Orientation dialog box. Click the
Isometric view 🔲 icon.

☀ Organize parts into folders. The folder for this chapter is
named SOLIDWORKS 2020. All documents for this book are saved in the
SOLIDWORKS 2020 folder.

☀ FeatureManager tabs and folder files will vary depending on system setup and
SOLIDWORKS Add-ins.

In the next section, edit the Sketch and Sketch plane.

Activity: Edit the Base Sketch and Sketch Plane

Edit the Base Sketch from 20mm to 40mm.

43) Right-click **Sketch1** in the FeatureManager.

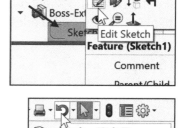

44) Click **Edit Sketch** ✎ from the Context toolbar.

45) Double-click **20** in the Graphics window. The Modify dialog box is displayed.

46) Enter **40**mm in the Modify dialog box.

47) Click the **Green Check mark** ✔ in the Modify dialog box. The diameter of the circle is 40mm. View the results.

Undo the modification from 20mm to 40mm.

48) Click **Undo** ↰ from the Main menu as illustrated.

49) Click **Exit Sketch** ⤴ from the Sketch toolbar to return to the original model.

Edit the Base Sketch Plane.

50) Right-click **Sketch1** in the FeatureManager.

51) Click **Edit Sketch Plane** 🔧 from the Context toolbar. The Sketch Plane PropertyManager is displayed.

52) **Expand** the fly-out Wheel FeatureManager as illustrated.

53) Click **Top Plane** from the fly-out Wheel FeatureManager. Top Plane is displayed in the Sketch Plane/Face box.

54) Click **OK** ✔ from the Sketch Plane PropertyManager.

55) **View** the model orientation in the Graphics window.

Return to the original orientation.

56) Click **Undo** ↰. The original orientation is displayed (Front Plane).

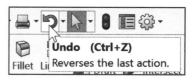

Display Modes, View Modes, View Tools, and Appearances

Access the display modes, view modes, view tools and appearances from the Standard Views toolbar and the Heads-up View toolbar. Apply these tools to display the required modes in your document.

The Apply scene tool adds the likeness of a material to a model in the Graphics window without adding the physical properties of the material. Select a scene from the drop-down menu.

Activity: Create the Second Feature of the Wheel - Revolved Boss

Create the second feature on the Right Plane.

57) Right-click **Right Plane** from the FeatureManager.

58) Click **Sketch** ⌐ from the Context toolbar.

Display a Right view and Zoom out.

59) Click **Right view** ⬚ from the Heads-up View toolbar.

60) Press the **z** key approximately five times to zoom out.

Display the origin.

61) If needed, click **View**, **Hide/Show**, **Origins** from the Main menu. The origin is displayed in the Graphics window.

Create the profile of the Wheel.

62) Click the **Sketch** tab from the CommandManager.

63) Click the **Centerline** Sketch tool from the drop-down menu as illustrated. The Insert Line PropertyManager is displayed.

64) Click the **midpoint** (it is very important that you select the midpoint) of the top edge of Boss-Extrude1 as illustrated. A Midpoint relation is displayed.

65)   Sketch a **vertical centerline** from the midpoint as illustrated. Approximately 65mm.

Deselect the Centerline Sketch tool.

66)   Right-click a **position** in the Graphics window. The Context toolbar is displayed.

67)   Click **Select** . The mouse pointer displays the Select icon and the Centerline Sketch tool is deactivated. The sketch is displayed in black.

Select the Line Sketch tool.

68)   Click the **Line** Sketch tool. The Insert Line PropertyManager is displayed.

69)   Click the **midpoint** of the top edge of Boss-Extrude1.

70)   Click a **position** to the right as illustrated. Approximately 5mm long. A horizontal relation is displayed.

71)   Click a **position** directly above as illustrated. Approximately 30mm long. A vertical relation is displayed.

72)   Sketch an **angle (*30degrees or LESS*) line** approximately 5mm long as illustrated. This is a very important step.

73)   Sketch a **vertical line** as illustrated. Approximately 5mm long.

74)   Sketch a **horizontal line** to complete the sketch.

75)   Click the **Centerline** to end the sketch. Note the automatic geometric relations in the sketch.

Deselect the Line Sketch tool.

76)   Right-click a **position** in the Graphics window. The Context toolbar is displayed.

77) Click **Select** . The mouse pointer displays the Select icon and the Centerline Sketch tool is deactivated.

In SOLIDWORKS, relations between sketch entities and model geometry, in either 2D or 3D sketches, are an important means of building in design intent. As you sketch, allow the SOLIDWORKS application to automatically add relations. Automatic relations rely on Inferencing, Pointer display, Sketch Snaps and Quick Snaps. After you sketch, manually add relations if needed and the dimensions to fully define the sketch.

To manually add a geometric relation, click a sketch entity. Hold the Ctrl key down. Click the second sketch entity. The Properties PropertyManager is displayed. Click the needed geometric relation. In this case, a Parallel relation was created between the centerline and the first vertical line.

Add relations, then dimensions. This keeps the user from having too many unnecessary dimensions. This also helps to show the design intent of the model.

Mirror the open sketch profile.

78) Click the **Mirror Entities** Sketch tool. The Mirror PropertyManager is displayed.

79) Click inside the **Mirror about** box.

80) Click the **Centerline** in the Graphics window. Line1 is displayed in the Mirror about box.

81) Click inside the **Entities to mirror** box.

Window-select the sketch entities.

82) Click in the **upper left corner** of the Graphics Window as illustrated.

83) **Drag** the mouse pointer to the lower right corner as illustrated.

84) **Release** the mouse button. The 5 sketch entities are displayed in the Entities to mirror box.

85) Click **OK** ✅ from the Mirror PropertyManager.

Deselect the Sketch Relations.
86) Click **View**, **Hide/Show**, uncheck **Sketch Relations** from the Main menu. View the results in the Graphics window.

Dimension the Sketch.
87) Click **Smart Dimension** ✎ from the Sketch toolbar.

The pointer displays the dimension symbol ✎ icon.

Dimension the angle.
88) Click the **two** illustrated lines. A dimension value is displayed.

89) Click a **position** to the left as illustrated. The Modify dialog box is displayed. The Smart Dimension tool uses the Smart Dimension PropertyManager. The PropertyManager provides the ability to select three tabs. Each tab provides a separate menu.

90) Enter **30**deg in the Modify dialog box. The Dimension Modify dialog box provides the ability to select a unit drop-down menu to modify units in a sketch or feature from the document properties.

91) Click the **Green Check mark** ✅ in the Modify dialog box.

92) **Zoom in** on the bottom horizontal line.

Dimension the bottom horizontal line.
93) Click the **bottom horizontal line** as illustrated.

94) Click a position **below** the model.

95) Enter **4**mm in the Modify dialog box.

96) Click the **Green Check mark** ✅ in the Modify dialog box.

Fit the model to the Graphics window.
97) Press the **f** key.

Dimension the overall width of the top horizontal line.
98) Click the **top horizontal line** as illustrated.

99) Click a position **above** the model.

100) Enter **10**mm in the Modify dialog box.

101) Click the **Green Check mark** in the Modify dialog box.

Dimension the first vertical line to the right.
102) Click the **first vertical line**.

103) Click a position to the **right** of the line.

104) Enter **30**mm in the Modify dialog box.

105) Click the **Green check mark** in the Modify dialog box.

Dimension the overall height.
106) Click the **top horizontal line**.

107) Click the **bottom horizontal line**.

108) Click a position to the **right** of the line as illustrated.

109) Enter **35**mm in the Modify dialog box.

110) Click the **Green check mark** in the Modify dialog box.
The sketch is fully defined and is displayed in black.

Display an Isometric view.

111) Click **Isometric view**
from the Heads-up View
toolbar.

Create a Revolved Boss Feature.

112) Click the **Features** tab.

113) Click **Revolved Boss/Base** from the Features toolbar. The Revolve PropertyManager is displayed.

114) Right-click inside the **Axis of Revolution** box.

115) Click **Delete**.

Display the Temporary Axes.

116) Click **View Temporary Axes** from the Hide/Show Items in the Heads-up toolbar.

117) Click the **Temporary Axis** in the Graphics window as illustrated. Axis 1 is displayed in the Axis of Revolution box.

118) Enter **360** for Direction 1 Angle.

119) Click **OK** from the Revolve PropertyManager. View the results in the FeatureManager and in the Graphics window. Revolve1 is created.

Deactivate the Temporary Axes.

120) Uncheck **View Temporary Axes** from the Hide/Show Items in the Heads-up toolbar.

121) Click **inside** the Graphics window.

122) Expand Revolve1 in the FeatureManager. Sketch2 is fully defined.

Activity: Create the Third Feature of the Wheel - Extruded Cut (Slot)

Display a Front view.

123) Click **Front view** from the Heads-up View toolbar.

Utilize the Centerpoint Straight Slot Sketch tool on the front face of the Wheel.

124) Right-click the **front face** of the Wheel as illustrated. The Context toolbar is displayed.

125) Click **Sketch** ⌐ from the Context toolbar.

126) Click **Centerpoint Straight Slot** ⊂⊃ from the Consolidated Sketch toolbar. The Slot PropertyManager is displayed. If needed, **uncheck** the Add dimensions box.

127) Click a **position** directly above the origin (first point).

128) Click a **position** directly above the first point (second point) as illustrated.

129) Click a **position** directly to the right of the second point (third point) as illustrated. The Centerpoint Straight Slot requires three points. Note: Add any needed geometric relations.

Utilize Construction geometry. Create a construction circle to locate the position of the center of the Slot relative to the Origin of the Wheel.

Locate the center of the slot. Create a Construction Circle.

130) Click the **Circle** ⊙ Sketch tool.

131) Click the **Origin** of the Wheel.

132) Click the **First point** of the slot.

133) Check the **For construction** box as illustrated.

134) Click **OK** ✔ from the Circle PropertyManager. A Construction circle is displayed.

Circle

Sketches a circle. Click to place the center of the circle, then drag and release to define its radius.

Insert dimensions to fully define the sketch.

135) Click **Smart Dimension** ✎ from the Sketch toolbar. The pointer displays the dimension symbol ✎ icon.

136) Enter the **dimensions** as illustrated. The sketch is fully defined and is displayed in black.

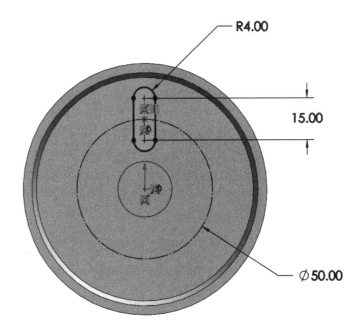

Create the Extruded Cut feature. The Extruded Cut feature is the Seed feature for the Circular Pattern. A Seed feature is the original feature for any type of pattern (linear, circular, sketch driven, curve driven, fill or mirrored pattern).

Create the Extruded Cut Feature.

137) Click the **Features** tab.

138) Click **Extruded Cut** 📄 from the Features toolbar. The Cut Extrude PropertyManager is displayed.

139) Select **Through All** for End Condition. View the End Condition options.

140) Click **OK** ✔ from the Cut-Extrude PropertyManager. View the results in the Graphics window. Cut-Extrude1 is highlighted in the FeatureManager.

💡 Use various end conditions to ensure proper design intent.

View Sketch3.
141) Expand Cut-Extrude1 in the Part FeatureManager. Sketch3 is fully defined.

Cut Extrude1 is the seed feature for the Circular Pattern of slots on the Wheel. In the next section, create the Circular Pattern feature.

Save the model.

142) Click **Save** 💾.

☀️ CommandManager and FeatureManager tabs and folder files will vary depending on system setup and Add-ins.

Select the Seed Feature for the Circular Pattern.
143) Click **Cut-Extrude1** in the FeatureManager. Cut-Extrude1 is highlighted.

Activity: Create the Fourth Feature of the Wheel - Circular Pattern

Display an Isometric view.

144) Click **Isometric view** from the Heads-up View toolbar.

Create the Circular Pattern feature.

145) Click **Circular Pattern** from the Features toolbar. The Circular Pattern PropertyManager is displayed. Cut-Extrude1 is selected and displayed in the Features to Pattern box.

Display the Temporary Axes.

146) Click **View Temporary Axes** from the Hide/Show Items in the Heads-up toolbar.

147) Click the **Center Temporary Axis** in the Graphics window. Axis <1> is displayed in the Pattern Axis box.

148) Check the **Equal spacing** box.

149) Enter **8** for Number of Instances.

150) Click **OK** ✔ from the Circular Pattern PropertyManager. CirPattern1 is displayed in the FeatureManager. View the circular pattern of slots in the model.

Deactivate the Temporary Axes.

151) Uncheck **View Temporary Axes** from the Hide/Show Items in the Heads-up toolbar.

Save the model.

152) Click **Save** 💾.

The Wheel requires a hole. Apply the Hole Wizard 🕳 feature. The Hole Wizard creates simple and complex Hole features by stepping through a series of options to define the hole type and hole position. The Hole Wizard requires a face or Sketch plane to position the Hole feature. Select the Front face of the hub.

Activity: Create the Fifth Feature of the Wheel - Hole

Display a Front view.

153) Click **Front view** 🔲 from the Heads-up View toolbar.

Insert the hole using the Hole Wizard tool.

154) Click the **Hole Wizard** 🕳 Feature tool. The Hole Specification PropertyManager is displayed.

155) Click the **Type** tab.

156) Click the **Hole** icon for Hole Type.

157) Select **ANSI Metric** for Standard.

158) Select **Drill sizes** for Type.

159) Select **Ø5.0** for Size.

160) Select **Through All** for End Condition. Accept the default settings.

161) Click the **Positions** tab.

Select the Sketch plane for the Hole Wizard.

162) Click the **front face** of Base-Extrude1 as illustrated. Do not click the origin.

The Point ▱ tool is selected.

163) Click the **origin** to place the center point of the hole. Right-click **Select** in the Graphics window to de-select the Point tool.

164) Click **OK** ✔ from the Hole Position PropertyManager. View the results.

Fit the WHEEL to the Graphics window.
165) Press the **f** key.

The Hole Wizard tool creates two sketches. The first sketch is the hole profile. The second sketch is to fully define the hole location relative to the origin.

Fillet/Round creates a rounded internal or external face on the part. You can fillet all edges of a face, selected sets of faces, selected edges, or edge loops.

In the next section, apply the Fillet feature to the Wheel. Create a Constant Size Fillet feature to 6 edges and 2 faces with a 2mm radius.

Activity: Create the Sixth Feature of the Wheel - Constant Size Fillet Feature

---

Create a Constant Size Fillet Feature on six edges and two faces.

166) Click the **Fillet** ⬭ Feature from the Features toolbar. The Fillet PropertyManager is displayed.

167) Click the **Manual** tab.

168) Click **Constant Size Fillet** for Fillet Type.

169) Enter **2**mm for Radius in the Fillet Parameters box.

170) Click the **front narrow face** of Revolve 1 as illustrated. Face1 is displayed in the Items To Fillet box.

171) Click the **front edge** of Boss-Extrude1 in the Graphics window. Edge1 is displayed in the Items To Fillet box.

172) Click the **back edge** of the hub. Edge2 is displayed in the Items To Fillet box.

173) Click the **inside circular edge** of Revolve1. Edge3 is displayed in the Items To Fillet box.

174) Rotate the **Wheel** and select the **three edges** and a **face** as above. Edge4, Edge5, Edge6 and Face2 are displayed in the Items To Fillet box. Zoom-in on the selected edges.

175) Click **OK** ✔ from the Fillet FeatureManager. Fillet1 is displayed in the FeatureManager.

176) Rotate the Wheel and view the results in the Graphics window.

Save the model.

177) Click **Save** .

Display a Trimetric view.

178) Click **Trimetric view** from the Heads-up View toolbar.

The Material dialog box helps you manage physical materials. You can work with pre-defined materials, create custom materials, apply materials to parts, and manage favorites.

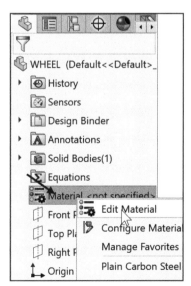

The left side of the Material dialog box contains a tree of available material types and materials. Tabs on the right display information about the selected material. If SOLIDWORKS Simulation is added in, more tabs appear.

In the next section, apply material (6061 Alloy) to the Wheel.

Activity: Apply Material to the Wheel

Apply Material.

179) Right-click **Material** from the FeatureManager.

180) Click **Edit Material**. The Material dialog box is displayed. View your options and material choices.

181) **Expand** the Aluminum Alloy folder.

182) Click **6061 Alloy**.

183) Click **Apply**.

184) Click **Close**. The Material is applied to the model.

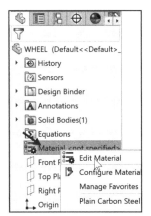

Use the new Search feature in the Material dialog box, to quickly locate material type.

Activity: View the Mass Properties of the Wheel

View the calculated mass properties of the
model. You can assign values for mass,
center of mass, and moments of inertia to
override the calculated values.

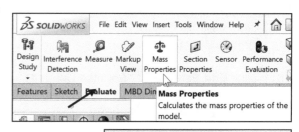

View the Mass Properties of the Wheel.
185) Click the **Evaluate** tab from the
CommandManager.

186) Click the **Mass Properties** icon.
The Mass Properties dialog box is
displayed. View the results. The total mass is 83.27
grams. The numbers represent the document properties
(2 decimal places). Your number may differ by 1- 2%
due to fillets.

Modify the precision in the Mass Properties dialog box to view
Density.
187) Click the **Options** button.

188) Check the **Use custom settings** button.

189) Select **4** for Decimal places.

190) Click **OK** from the Mass/Section Property Options dialog
box. Density = .0027 grams per cubic millimeters.

Close the Mass Properties dialog box.
191) Click **Close** ⊠.

Display an Isometric view - Shaded With Edges.
192) Press the **space bar** to display the Orientation dialog
box.

193) Click the **Isometric view** 🔲 icon. You can also access
the Isometric view tool from the Heads-up View toolbar.

194) Click the **Shaded With Edges** 🔲 icon.

Save the Wheel.
195) Click **Save** 💾.

In the next section, modify the Circular Pattern feature.
Decrease the number of instances from 8 to 4. Compare
the Mass Properties of the modified model.

Mass properties of Wheel
   Configuration: Default
   Coordinate system: -- default --

Density = 0.00 grams per cubic millimeter

Mass = 83.27 grams

Volume = 30842.24 cubic millimeters

Surface area = 15835.01 square millimeters

Center of mass: (millimeters)
   X = 0.00
   Y = 0.00
   Z = 0.00

Activity: Modify the Circular Pattern. Decrease the Number of Instances.

Modify the CirPattern1 feature.

196) Right-click **CirPattern1** from the FeatureManager.

197) Click **Edit Feature** from the Content toolbar. The CirPattern1 PropertyManager is displayed.

198) Enter **4** for Number of Instances.

199) Click **OK** from the CirPattern PropertyManager. View the results in the PropertyManager.

Activity: View the New Mass Properties of the Wheel

200) Click the **Evaluate** tab from the CommandManager.

201) Click the **Mass Properties** icon. The Mass Properties dialog box is displayed. View the results. The total mass is 90.63 grams vs. 83.27 grams. You removed 4 slots (instances) from the model.

Close the Mass Properties dialog box.
202) Click **Close**.

Activity: Modify the Circular Pattern. Return to the Original Instances.

Modify the CirPattern1 feature.

203) Right-click **CirPattern1** from the FeatureManager.

204) Click **Edit Feature** from the Content toolbar. The CirPattern1 PropertyManager is displayed.

205) Enter **8** for Number of Instances.

206) Click **OK** from the CirPattern PropertyManager. View the results in the PropertyManager.

In the next section, add an Appearance to the Wheel. Use the Appearances PropertyManager to apply colors, material appearances, and transparency to parts and assembly components. The Appearance tool does not modify the mass properties.

Activity: Add an Appearance to the Wheel

Add an Appearance to the Wheel.

207) Right-click **Wheel** in the FeatureManager.

208) Click the **Appearances** drop-down arrow.

209) Click the **WHEEL** icon as illustrated. The color PropertyManager is displayed.

210) Click a **color (yellow)** from the Color dialog box.

211) Click **OK** ✔ from the color PropertyManager. View the results in the Graphics window.

Display an Isometric view - Shaded With Edges.

212) Press the **space bar** to display the Orientation dialog box.

213) Click the **Isometric view** 🔲 icon. You can also access the Isometric view tool from the Heads-up View toolbar.

214) Click the **Shaded With Edges** 🔲 icon.

Save the Wheel.

215) Click **Save** 💾. You are finished with the part.

You created a new part (Wheel) that is used in the next chapter to create the Fly Wheel sub-assembly. The Fly Wheel sub-assembly with other provided components is used to create the final Stirling Engine assembly.

Summary

You established a SOLIDWORKS session and created a new part (Wheel) with user defined document properties. The correct Sketch plane and orientation for the Wheel was selected.

The Wheel part is used in the Fly Wheel sub-assembly. The sub-assembly is used in the final Stirling Engine assembly.

The following sketch tools were utilized: Circle, Line, Centerline, Center point Straight Slot and Mirror Entities.

The following features were utilized: Extruded Boss/Base, Extruded Cut, Revolved Boss/Boss, Hole Wizard, Circular Pattern and Fillet.

Material and Appearance were applied with the proper design intent.

During the creation of the Wheel you learned about various Sketch states, editing Sketches and Features along with applying the following geometric relations: Coincident, Vertical, Horizontal, Tangent and Midpoint.

If you modify a document property from an Overall drafting standard, a modify message is displayed as illustrated.

Overall drafting standard
ANSI-MODIFIED
Derived from: ANSI

The FeatureManager consists of various tabs:

- *FeatureManager design tree* tab.

- *PropertyManager* tab.

- *ConfigurationManager* tab.

- *DimXpertManager* tab.

- *DisplayManager* tab.

- *CAM FeatureManager tree* tab.

- *CAM Operation tree* tab.

- *CAM Tools tree* tab.

Click the direction arrows to expand or collapse the FeatureManager design tree.

CommandManager and FeatureManager tabs and tree folders will vary depending on system set-up and SOLIDWORKS Add-ins.

Use the new Search feature in the Material dialog box, to quickly locate material type.

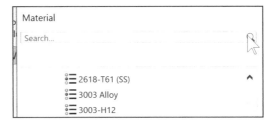

In Chapter 3, establish a SOLIDWORKS session and create two new assemblies with user defined document properties:

- Fly Wheel

- Stirling Engine

Insert the following Standard and Quick mate types: Coincident, Concentric, Distance and Tangent.

Utilize the following assembly tools: Insert Component, Suppress, Un-suppress, Mate, Move Component, Rotate Component, Interference Detection, Hide, Show, Flexible, Ridge, and Multiple mate mode.

Create an Exploded View with animation.

Apply the Measure and Mass Properties tool to modify a component in the Stirling Engine assembly.

Save the final Stirling Engine assembly, using the SOLIDWORKS Pack & Go tool.

Questions

1. Explain the steps in starting a SOLIDWORKS session.

2. Describe the procedure to begin a new 2D sketch.

3. Explain the steps required to modify units in a part document from inches to millimeters.

4. Describe the procedure to create a simple 3D cylinder on the Front Plane with an Extruded Boss/Base (Boss-Extrude1) feature.

5. Identify the three default Reference planes in SOLIDWORKS.

6. Describe a Base feature. Provide two examples from this chapter.

7. Describe the differences between an Extruded Boss/Base feature and an Extruded Cut feature.

8. The sketch color black indicates a sketch is _____ defined.

9. The sketch color blue indicates a sketch is _____ defined.

10. The sketch color red indicates a sketch is _____ defined.

11. Describe the procedure to "wake up" a center point in a sketch.

12. Define a Geometric relation. Provide three examples.

13. Describe the procedure to create a Linear Pattern feature.

14. Describe an assembly or sub-assembly.

15. What are mates and why are they important in assembling components?

16. In an assembly, each component has_____# degrees of freedom. Name them.

17. True or False. A fixed component cannot move in an assembly.

18. Review the Design Intent section in the book. Identify how you incorporated design intent into a part or assembly document.

Exercises

Exercise 2.1: Identify the Sketch plane for the Boss-Extrude1 (Base) feature as illustrated. Simplify the number of features.

A: Top Plane

B: Front Plane

C: Right Plane

D: Left Plane

Correct answer _____.

Create the part. Dimensions are arbitrary.

Origin

Exercise 2.2: Identify the Sketch plane for the Boss-Extrude1 (Base) feature as illustrated. Simplify the number of features.

A: Top Plane

B: Front Plane

C: Right Plane

D: Left Plane

Correct answer _____.

Create the part. Dimensions are arbitrary.

Origin

Exercise 2.3: Identify the Sketch plane for the Boss-Extrude1 (Base) feature as illustrated. Simplify the number of features.

A: Top Plane

B: Front Plane

C: Right Plane

D: Left Plane

Correct answer _____.

Create the part. Dimensions are arbitrary.

Origin

Exercise 2.4: FLATBAR - 3HOLE Part

Create an ANSI, IPS FLATBAR - 3HOLE part document.

- Utilize the Front Plane for the Sketch plane. Insert an Extruded Base (Boss-Extrude1) feature. Do not display Tangent Edges.

- Create an Extruded Cut feature. This is your seed feature. Apply the Linear Pattern feature. The FLATBAR - 3HOLE part is manufactured from 0.06in., [1.5mm] 6061 Alloy.

Exercise 2.5: FLATBAR - 5HOLE Part

Create an ANSI, IPS, FLATBAR - 5HOLE part as illustrated.

- Utilize the Front Plane for the Sketch plane. Insert an Extruded Base (Boss-Extrude1) feature.

- Create an Extruded Cut feature. This is your seed feature. Apply the Linear Pattern feature. The FLATBAR - 5HOLE part is manufactured from 0.06in, [1.5mm] 6061 Alloy.

- Calculate the required dimensions for the FLATBAR - 5HOLE part. Use the following information: Holes are .500in. on center, Radius is .250in., and hole diameter is .190in.

- Do not display Tangent edges in the final model.

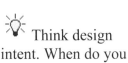 Think design intent. When do you use the various End Conditions and Geometric sketch relations? What are you trying to do with the design? How does the component fit into an assembly?

Exercise 2.6: Simple Block Part

Create the illustrated ANSI part. Note the location of the Origin in the illustration.

- Calculate the overall mass of the illustrated model.

- Apply the Mass Properties tool.

- Think about the steps that you would use to build the model.

- Review the provided information carefully.

- Units are represented in the IPS (inch, pound, second) system.

- A = 3.50in, B = .70in

Given:
A = 3.50
B = .70
Material: 1060 Alloy
Density = 0.0975 lb/in^3
Units: IPS
Decimal places = 2

Origin

Exercise 2.7: Simple Block Part

Create the illustrated ANSI part. Note the location of the Origin in the illustration.

Create the sketch symmetric about the Front Plane. The Front Plane in this problem is **not** your Sketch Plane. Utilize the Blind End Condition in Direction 1.

Given:
A = 3.00
B = .75
Material: Copper
Density = 0.321 lb/in^3
Units: IPS
Decimal places = 2

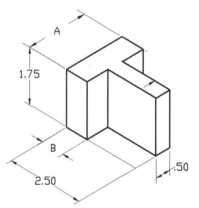

- Calculate the overall mass of the illustrated model.

- Apply the Mass Properties tool.

- Think about the steps that you would use to build the model.

- Review the provided information carefully. Units are represented in the IPS (inch, pound, second) system.

- A = 3.00in, B = .75in

Note: Sketch1 is symmetrical.

Origin

Exercise 2.8: Simple Block Part

Create an ANSI part from the illustrated model. Note the location of the Origin in the illustration.

- Calculate the volume of the part and locate the Center of mass with the provided information.

- Apply the Mass Properties tool.

- Think about the steps that you would use to build the model.

- Review the provided information carefully.

Given:
A = 3.30
B = 2.00
Material: 2014 Alloy
Density = .101 lb/in^3
Units: IPS
Decimal places = 2

Origin

Exercise 2.9: Simple Block Part

Create an ANSI, MMGS part from the illustrated drawing: Front, Top, Right and Isometric views.

Note the location of the Origin in the illustration. The drawing views are displayed in Third Angle Projection.

- Apply 1060 Alloy for material.

- Calculate the Volume of the part.

- Locate the Center of mass.

Think about the steps that you would use to build the model. The part is symmetric about the Front Plane.

Origin

Exercise 2.10: Simple Block Part

Create the ANSI, MMGS part from the illustrated drawing: Front, Top, Right and Isometric views.

Origin

- Apply 1060 Alloy for material.

- The part is symmetric about the Front Plane.

- Calculate the Volume of the part and locate the Center of mass.

Think about the steps that you would use to build the model.

The drawing views are displayed in Third Angle Projection.

Exercise 2.11: LINKAGE-2 Assembly

Create the LINKAGE-2 assembly.

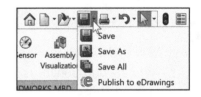

- Copy the LINKAGE assembly from the Chapter 2 Homework folder to your local hard drive.

- Open the assembly.

- Select Save As from the drop-down Menu bar.

- Check the Save as copy and open check box.

- Enter LINKAGE-2 for file name. LINKAGE-2 ASSEMBLY for description.

File name: LINKAGE-2
Save as type: SOLIDWORKS Assembly (*.asm;*.sldasm)
Description: LINKAGE-2 ASSEMBLY

- Save as
- Save as copy and continue
- Save as copy and open
- Include all referenced components
- Add prefix
- Add suffix   Advanced
- Hide Folders

The FLATBAR-3HOLE part was
created in Exercise 2.4. Utilize two
AXLE parts, four SHAFT COLLAR
parts, and two FLATBAR-3HOLE
parts to create the LINKAGE-2
assembly as illustrated. The AXLE and
SHAFT COLLAR is provided in the
Homework folder.

- Insert the first AXLE part.

- Insert a Concentric mate.

- Insert a Coincident mate.

- Insert the first FLATBAR-3HOLE part.

- Insert a Concentric mate.

- Insert a Coincident mate.

- Perform the same procedure for the second
  FLATBAR-3HOLE part.

- Insert a Parallel mate between the 2 FLATBAR-
  3HOLE parts. Note: The 2 FLATBAR-3HOLE
  parts move together.

When a component is in the Lightweight
state, only a subset of its model data is loaded into
memory. The remaining model data is loaded on an
as-needed basis.

When a component is fully resolved, all its
model data is loaded into memory.

- Insert the second AXLE part.

- Insert a Concentric mate.

- Insert a Coincident mate.

- Insert the first SHAFT-COLLAR part.

- Insert a Concentric mate.

- Insert a Coincident mate.

- Perform the same tasks to insert the other three required SHAFT-COLLAR parts as illustrated.

Exercise 2.12: LINKAGE-2 Assembly Motion Study

Create a Motion Study using the LINKAGE-2 Assembly that was created in the previous exercise.

- Create a Basic Motion Study.

- Apply a Rotary Motor to the front FLATBAR-3HOLE as illustrated.

- Play and Save the Simulation.

Exercise 2.13: ROCKER Assembly

Create a ROCKER assembly. The ROCKER assembly consists of two AXLE parts, two FLATBAR-5HOLE parts, and two FLATBAR-3HOLE parts.

The FLATBAR-3HOLE parts are linked together with the FLATBAR-5HOLE.

The three parts rotate clockwise and counterclockwise, above the Top Plane. Create the ROCKER assembly.

- Insert the first FLATBAR-5HOLE part. The FLATBAR-5HOLE is fixed to the Origin of the ROCKER assembly.

- Insert the first AXLE part.

- Insert a Concentric mate.

- Insert a Coincident mate.

- Insert the second AXLE part.

- Insert a Concentric mate.

- Insert a Coincident mate.

- Insert the first FLATBAR-3HOLE part.

- Insert a Concentric mate.

- Insert a Coincident mate.

- Insert the second FLATBAR-3HOLE part.

- Insert a Concentric mate.

- Insert a Coincident mate.

- Insert the second FLATBAR-5HOLE part.

- Insert the required mates.

Note: The end holes of the second FLATBAR-5HOLE are concentric with the end holes of the FLATBAR-3HOLE parts.

Note: In mechanical design, the ROCKER assembly is classified as a mechanism. A Four-Bar Linkage is a common mechanism comprised of four links.

Link1 is called the Frame.

The AXLE part is Link1.

Link2 and Link4 are called the Cranks.

The FLATBAR-3HOLE parts are Link2 and Link4. Link3 is called the Coupler. The FLATBAR-5HOLE part is Link3.

If an assembly or component is loaded in a Lightweight state, right-click the assembly name or component name from the FeatureManager. Click Set Lightweight to Resolved.

Determine the static and dynamic behavior of mates in each sub-assembly before creating the top level assembly.

Exercise 2.14: 4 Bar linkage

Create the 4 bar linkage assembly as illustrated. Create the five components. Assume dimensions.

In an assembly, fix (f) the first component to the origin or fully define it to the three default reference planes.

Insert all needed mates to simulate the movement of a 4 bar linkage assembly.

Read the section on Coincident, Concentric and Distance mates in SOLIDWORKS help.

Create a base with text for extra credit.

Below are a few sample models from my Freshman Engineering class.

Note the different designs to maintain the proper movement of a 4 bar linkage using a base.

Exercise 2.15: Fill Pattern

Create a Polygon Layout Fill Pattern feature. Apply the seed cut option.

1. Open **Fill Pattern 2-1** from the Chapter 2 Homework folder.

2. Click the **Fill Pattern** 🖼 Features tool. The Fill Pattern PropertyManager is displayed.

3. Click the **Front face** of Extrude1. Face<1> is displayed in the Fill Boundary box. The direction arrow points to the right.

4. Click **Polygon** for Pattern Layout.

5. Click **Target spacing**.

6. Enter **15**mm for Loop Spacing.

7. Enter **6** for Polygon sides.

8. Enter **10**mm for Instances Spacing.

9. Enter **10**mm for Margins. View the direction arrow.

10. Click **Create seed cut**.

11. Click **Circle** for Features to Pattern.

12. Enter **4mm** for Diameter.

13. Click **OK** ✔ from the Fill Pattern PropertyManager. Fill Pattern1 is created and is displayed in the FeatureManager.

14. **View** the results.

Exercise 2.16: Fill Pattern

Create a Circular Layout Fill Pattern feature.

1. **Copy** the Chapter 2 Homework folder to your hard drive.

2. Open **Fill Pattern 2-2** from the Chapter 2 Homework folder.

3. Click the **Fill Pattern** Features tool. The Fill Pattern PropertyManager is displayed.

4. Click the **top face** of Boss-Extrude1. Face<1> is displayed in the Fill Boundary box. The direction arrow points to the back. Click **Circular** for Pattern Layout.

5. Enter **.10**in for Loop Spacing.

6. Click **Target spacing**.

7. Enter .**10**in Instance Spacing.

8. Enter **.05**in for Margins. Edge<1> is selected for Pattern Direction.

9. Click **inside** the Features to Pattern box.

10. Click **Boss-Extrude2** from the fly-out FeatureManager. Boss-Extrude2 is displayed in the Features to Pattern box.

11. Click **OK** from the Fill Pattern PropertyManager. Fill Pattern1 is created and is displayed in the FeatureManager.

12. **View** the results.

Exercise 2.17: Tangent Mate

Create a Tangent mate between the roller-wheel and the cam on both sides to simulate movement per the avi file.

1. **Copy** the Chapter 2 Homework folder to your hard drive.

2. Open **Tangent** from the Chapter 2 Homework\Tangent folder. The assembly is displayed.

Create a Tangent mate between the roller-wheel and the cam on both sides. The cam was created with imported geometry (curve).

A Tangent mate places the selected items tangent to each other. At least one selected item must be either a conical, cylindrical, spherical face. An Angle mate places the selected items at the specified angle to each other.

3. Click the **Mate** tool from the Assembly tab.

4. **Pin** the Mate PropertyManager.

Create the first Tangent mate.

5. Click the **contact face** of the left cam as illustrated.

6. Click the **contact face** of the left roller-wheel as illustrated. Click **Tangent** mate from the PropertyManager. Click **OK** from the Tangent1 PropertyManager.

You can set an option in the Mate PropertyManager so that the first component you select from becomes transparent. Then selecting from the second component is easier, especially if the second component is behind the first.

Create the second Tangent mate.

7. Click the **contact face** of the right cam. Click the **contact face** of the right roller-wheel.

8. Click **Tangent** mate from the PropertyManager. Click **OK** from the Tangent2 PropertyManager.

9. **Un-Pin** the Mate PropertyManager. Click **OK** from the Mate PropertyManager. Display an **Isometric view**.

10. **Move** the cam. View the results.

Exercise 2.18: Motion Study 1

Apply a motor to move components in an assembly. Set up an animation motion study. Use the motion study time line and the MotionManager design tree to suppress mates in the motion study.

1. Copy the **Chapter 2 Homework folder** to your hard drive.

2. Open **Motion Study 1** from the Chapter 2 Homework\Motion Study folder.

3. Click the **Motion Study 1** tab at the bottom of the Graphics window.

4. Select **Basic Motion** for Study type.

5. Click the **Motor** 🔧 tool from the MotionManager toolbar. The Motor PropertyManager is displayed.

6. Click **Rotary Motor** for Motor Type.

7. Select **Face<1>@Part2-1** for Motor Direction in the Graphics window as illustrated. The direction arrow points counterclockwise.

8. Select **Constant speed** for Motor Type.

9. Enter **30** for Constant Speed Motor.

10. Click **OK** ✔ from the Motor PropertyManager.

Set the duration and run your animation on the 4bar-Linkage.

11. Drag the key to **6** seconds as illustrated.

12. Click **Play from Start** ▶ from the MotionManager. View the results.

In the next section, suppress a Concentric mate on a component. View the results in the Motion Study.

13. **Expand** the Mate folder in the MotionManager Design tree.

14. Right-click at 2 seconds and select **Place Key** 🔹 for the Concentric2 mate.

15. Right-click at 4 seconds and select **Place Key** 🔹 for the Concentric2 mate.

16. Set the time bar to **2 seconds**.

17. Right-click **Concentric2** in the MotionManager design tree.

18. Click **Suppress**. The mate is suppressed between the 2 and 4 second marks.

19. Click the **Calculate** 🗔 tool. After calculating a motion study, you can view it again.

20. Click **Play from Start** ▮▶ from the MotionManager toolbar. View the results. Explore your options and the available tools. View SOLIDWORKS Help for additional information on a Motion Study.

Return to the FeatureManager.

21. Click the **Model** tab to return to SOLIDWORKS.

22. **Close** the model.

Notes:

Notes:

Chapter 3

Assembly Modeling - Bottom up method

Below are the desired outcomes and usage competencies based on the completion of Chapter 3.

Desired Outcomes:	Usage Competencies:

• Assembly documents:
 	
  • Set drafting standard.
 	
  • Set units and precision.
 	
• Sub-assembly:
 	
  • Fly Wheel.
 	
• Final assembly:
 	
  • Stirling Engine
 	
• Exploded Sub-assembly.
 	
• Save An Assembly Document.
	• Set Document Properties as they relate to an Assembly.
	
	• Comprehend the assembly process. Insert parts/sub-components into an assembly.
	
	• Ability to apply Standard mates and Quick mates (Coincident, Concentric, Distance, and Tangent) in an assembly using the Bottom-up design approach.
	
	• Utilize the following assembly tools: Insert Component, Mate, Hide, Show, Move, Rotate, Modify, Flexible, Ridge, and Multiple mate mode.
	
	• Apply the Measure and Mass Properties tool.
	
	• Incorporate design changes into an assembly.

Notes:

Notes:

Chapter 3 - Assembly Modeling - Bottom up method

Chapter Objective

Create two assemblies. The first assembly is the Fly Wheel. The second assembly is the Stirling Engine. The Fly Wheel is a sub-assembly of the final Stirling Engine assembly.

On the completion of this chapter, you will be able to:

- Create a new assembly document.

- Set user defined document properties.

- Understand assembly modeling approach.

- Comprehend linear and rotational motion.

- Insert parts and sub-assemblies in an assembly.

- Insert and edit the following Standard mates:

  o Coincident, Concentric, Distance and Tangent.

- Incorporate design changes in an assembly.

- Edit and Modify features and components in an assembly.

- Utilize the following assembly tools: Insert Component, Mate, Hide, Show, Move, Rotate, Modify, Flexible, Ridge, and Multiple mate mode.

- Create an Exploded View with animation.

- Apply the Measure and Mass Properties tool.

- Apply the Pack and Go tool.

Activity: Start a SOLIDWORKS Session. Create a New Assembly.

Start a SOLIDWORKS session.

1) Double-click the **SOLIDWORKS 2020 icon** from the desktop.

2) **Close** the SOLIDWORKS Welcome dialog box.

3) **Pin** ⚲ the Menu bar as illustrated. Use the Menu bar and the Menu bar toolbar.

Create a New Assembly Document.

4) Click **New** ⬜ from the Menu bar or click **File**, **New** from the Menu bar menu. The New SOLIDWORKS Document dialog box is displayed. Note: Advanced mode is used in the book.

5)   Click the **Templates** tab.

6)   Double-click **Assembly** from the New SOLIDWORKS Document dialog box. The Begin Assembly PropertyManager and the Window Open dialog box is displayed.

7)   Click **Cancel** in the Open dialog box.

8)   Click **Cancel** ✖ from the Assembly FeatureManager. Assem1 is displayed in the FeatureManager.

The first system default Assembly filename is Assem1. The system attaches the .sldasm to the created assembly. The second created assembly in the same session increments to the filename Assem2.

The Begin Assembly PropertyManager and the Insert Component PropertyManager is displayed when a new or existing assembly is opened, if the Start command when creating new assembly box is checked.

Display the origin.

9)   Click **View**, **Hide/Show**, **Origins** from the Main menu. The origin is displayed in the Graphics window.

Directional input refers by default to the global coordinate system (X, Y and Z), which is based on Plane1 with its origin located at the origin of the part or assembly. Plane1 (Front) is the first plane that appears in the FeatureManager design tree and can have a different name. The reference triad shows the global X-, Y- and Z-directions.

Activity: Set Document Properties for the Fly Wheel Assembly

10)   Click **Options** ⚙ from the Menu bar. The System Options General dialog box is displayed.

11)   Click the **Document Properties** tab.

12)   Select **ANSI** from the Overall drafting standard drop-down menu. Various Detailing options are available depending on the selected standard.

🔅 Various detailing options are available depending on the selected standard.

The Overall drafting standard determines the display of dimension text, arrows, symbols, and spacing. Units are the measurement of physical quantities. Millimeter dimensioning and decimal inch dimensioning are the two most common unit types specified for engineering parts and drawings.

Set Document Properties. Enter units and precision.

13)   Click the **Units** folder.

14)   Click **MMGS** (millimeter, gram, second) for Unit system.

15)   Select **.12,** (two decimal places) for Length basic units.

16)   Click **OK** from the Document Properties - Units dialog box. The Assembly FeatureManager is displayed.

🔅 The Origin ⊥ represents the intersection of the Front, Top and Right Planes.

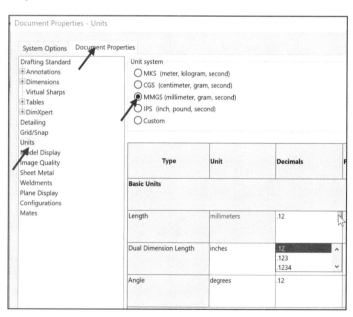

Assembly Modeling Approach

In SOLIDWORKS, components and their assemblies are directly related through a common file structure. Changes in the components directly affect the assembly and vice versa. You can create assemblies using the Bottom-up assembly approach, Top-down assembly approach or a combination of both methods. This chapter focuses on the Bottom-up assembly approach.

The Bottom-up approach is the traditional method that combines individual components. Based on design criteria, the components are developed independently. The three major steps in a Bottom-up assembly approach are create each component independent of any other component in the assembly, insert the components into the assembly, and mate the components in the assembly as they relate to the physical constraints of your design.

In the Top-down assembly approach, major design requirements are translated into layout sketches, assemblies, sub-assemblies and components.

In the Top-down approach, you do not need all of the required component design details. Individual relationships are required.

Example: A computer. The inside of a computer can be divided into individual key sub-assemblies such as a motherboard, disk drive, power supply, etc. Relationships between these sub-assemblies must be maintained for proper fit.

Linear Motion and Rotational Motion

In dynamics, motion of an object is described in linear and rotational terms. Components possess linear motion along the x, y and z-axes and rotational motion around the x, y and z-axes. In an assembly, each component has *six degrees* of freedom: three translational (linear) and three rotational. Mates remove degrees of freedom. All components are rigid bodies. The components do not flex or deform.

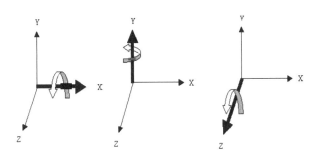

Activity: Insert the First Component into the Fly Wheel Assembly

The first component is the foundation of the assembly. The Bracket is the first component in the Fly Wheel assembly. The Bushing is the second component in the Fly Wheel assembly. All needed components for this chapter including the Wheel are in the SOLIDWORKS 2020\FLY WHEEL folder.

Add components to assemblies utilizing the following techniques:

- Utilize the Insert Components Assembly tool.

- Utilize Insert, Component from the Menu bar.

- Drag a component from Windows Explorer (3D ContentCentral) into the Assembly.

- Drag a component from the SOLIDWORKS Design Library into the Assembly.

- Drag a component from an Open part file into the Assembly.

Insert the Bracket component.

17) Click **Insert Component** from the Assembly toolbar. The Insert Component PropertyManager and the Window Open dialog box is displayed.

18) Double-click the **Bracket** part from the SOLIDWORKS 2020\FLY WHEEL folder.

19) Click **OK** from the Insert Component PropertyManager. The Bracket is fixed to the origin. It cannot translate or rotate.

The book is designed to expose the SOLIDWORKS user to numerous tools and procedures. It may not always use the simplest and most direct process.

To fix the first component to the Origin, click OK ✔ from the Begin Assembly PropertyManager or click the Origin in the Graphics window.

To remove the fixed state (f), right-click the **fixed component name** in the FeatureManager. Click **Float**. The component is free to move.

CommandManager and FeatureManager tabs and folder files will vary depending on system setup and Add-ins.

💡 Display an Isometric view of the model. Press the **space bar** to display the Orientation dialog box. Click the **Isometric view** 🧊 icon.

Save the Assembly.

20) If needed, deactivate the Origin. Click **View**, **Hide/Show** from the Main Menu. Click the **Origins** button.

21) Click **Save As** 🖼 from the Drop-down Menu bar.

22) Browse to the **SOLIDWORKS 2020** folder.

23) Enter **Fly Wheel** for the File name.

24) Enter **Fly Wheel** for the Description.

25) Click **Save** from the Save As dialog box. The Fly Wheel Assembly FeatureManager is displayed.

💡 Organize parts, assemblies and drawings into folders. All documents for this book are saved in the SOLIDWORKS folder.

Mate Types

Mates provide the ability to create geometric relationships between assembly components. Mates define the allowable directions of rotational or linear motion of the components in the assembly. Move a component within its degrees of freedom in the Graphics window to view the behavior of an assembly.

Mates are solved together as a system. The order in which you add mates does not matter. All mates are solved at the same time. You can suppress mates just as you can suppress features.

The Mate PropertyManager provides the ability to select either the **Mates** or **Analysis** tab. Each tab has a separate menu. The **Analysis** tab requires the ability to run SOLIDWORKS Motion. The Analysis tab is not covered in this book. The Mate PropertyManager displays the appropriate selections based on the type of mate you create. The components in the Fly Wheel assembly utilize Standard Mate types.

Review **Standard**, **Advanced** and **Mechanical** Mates types.

Standard Mates:

Components are assembled with various mate types. These are the Standard Mate types:

- **Coincident mate**. Locate the selected faces, edges, points or planes so they use the same infinite line. A Coincident mate positions two vertices for contact.

- **Parallel mate**. Locate the selected items to lie in the same direction and to remain a constant distance apart. A parallel mate permits only translational motion of a single part with respect to another. No rotation is allowed.

- **Perpendicular mate**. Locate the selected items at a 90-degree angle to each other. The perpendicular mate allows both translational and rotational motion of one part with respect to another.

You can apply a Perpendicular mate between a complex, non-analytic surface and a linear entity like a line, an edge, an axis, or an axial entity.

- **Tangent mate**. Place the selected items tangent to each other. At least one selected item must be either a conical, cylindrical, spherical face.

- **Concentric mate**. Locate the selected items so they can share the same center point. You can prevent the rotation of components that are mated with concentric mates by selecting the Lock Rotation option. Locked concentric mates are indicated by an icon in the FeatureManager design tree.

- **Lock mate**. Enable two components to be fully constrained to each other without needing to create a rigid sub-assembly. A real-world example of a lock mate is a weld that holds two parts together.

- **Distance mate**. Locate the selected items with a specified distance between them. Use the drop-down arrow box or enter the distance value directly.

- **Angle mate**. Locate the selected items at the specified angle to each other. Use the drop-down arrow box or enter the angle value directly. You can add a reference direction to help prevent angle mates from flipping unexpectedly.

- **Mate alignment**. Toggle the mate alignment as necessary.

- **Aligned**. Locate the components so the normal or axis vectors for the selected faces point in the same direction.

- **Anti-Aligned**. Locate the components so the normal or axis vectors for the selected faces point in the opposite direction.

Advanced Mates:

These are the Advanced Mate types:

- **Profile Center**. Mate to center automatically center-aligns common component types such as rectangular and circular profiles to each other.

- **Symmetric**. Force two similar entities to be symmetric about a planar face or plane.

- **Width**. Center a tab within the width of a groove. The Width mate is used to replace the Symmetric mate where components have tolerance and a gap rather than a tight fit.

- **Path Mate**. Enable any point on a component to be set to follow a defined path.

- **Linear/Linear Coupler**. Establish a relationship between the translation of one component and the translation of another component.

- **Limit**. Allow components to move within a range of values for distance and angle. Select the angle and distance from the provided boxes. Specify a starting distance or angle as well as a maximum and minimum value.

- **Distance**. Locate the selected items with a specified distance between them. Use the drop-down arrow box or enter the distance value directly (Limit mate).

- **Angle**. Locate the selected items at the specified angle to each other. Use the drop-down arrow box or enter the angle value directly.

- **Mate alignment**. Toggle the mate alignment as necessary. The options are:

  - **Aligned**. Locate the components so the normal or axis vectors for the selected faces point in the same direction.

  - **Anti-Aligned**: Locate the components so the normal or axis vectors for the selected faces point in the opposite direction.

Mechanical Mates:

These are the Mechanical Mate types:

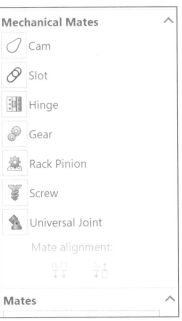

- **Cam**. Force a plane, cylinder, or point to be tangent or coincident to a series of tangent extruded faces. Four conditions can exist with the Cam mate: *Coincident, Tangent, CamMateCoincident* and *CamMateTangent*. Create the profile of the cam from lines, arcs and splines as long as they are tangent and form a closed loop.

- **Slot**. Constrain the movement of a bolt or a slot to within a slot hole. You can mate bolts to straight or arced slots and you can mate slots to slots. You can select an axis, cylindrical face, or a slot to create slot mates.

- **Hinge**. Limit the movement between two components to one rotational degree of freedom. It has the same effect as adding a Concentric mate plus a Coincident mate. You can also limit the angular movement between the two components.

- **Gear**. Force two components to rotate relative to one another around selected axes. The Gear mate provides the ability to establish gear type relations between components without making the components physically mesh. Note: SOLIDWORKS provides the ability to modify the gear ratio without changing the size of the gears. Align the components before adding the Mechanical gear mate.

- **Rack and Pinion**. Linear translation of one component (the rack) causes circular rotation in another component (the pinion), and vice versa. You can mate any two components to have this type of movement relative to each other. The components do not need to have gear teeth.

- **Screw**. Constrains two components to be concentric, and adds a pitch relationship between the rotation of one component and the translation of the other. Translation of one component along the axis causes rotation of the other component according to the pitch relationship. Likewise, rotation of one component causes translation of the other component. Align the components before adding the Mechanical screw mate.

- **Universal Joint**. Permit the transfer of rotations from one rigid body to another. This mate is particularly useful to transfer rotational motion around corners, or to transfer rotational motion between two connected shafts that are permitted to bend at the connection (drive shaft on an automobile).

🔆 Mates reflect the physical behavior of a component in an assembly. In this project, the two most common Mate types are Concentric and Coincident.

Quick Mate:

Quick Mate is a procedure to mate components together. No command (click Mate from the Assembly CommandManager) is required. Hold the Ctrl key down, and make your selections. Release the Ctrl key, the Quick Mate pop-up is displayed below the context toolbar. Select the mate and you are finished.

Utilize the Quick Mate procedure for Standard mates, Cam mate, Slot mate, Symmetric mate, Profile Center, Advance Distance and Angle mates. To activate the Quick Mate functionality, click Tools, Customize. On the toolbars tab, under Context toolbar settings, select Show Quick Mates. Quick Mate is selected by default.

Activity: Insert and Mate the Second Component - Bushing

Insert the Bushing part into the Fly Wheel Assembly. Deactivate the origins.

26) Click **Insert Component** from the Assembly toolbar. The Insert Component PropertyManager and the Window Open dialog box is displayed.

27) Double-click the **Bushing** part from the SOLIDWORKS 2020\FLY WHEEL folder.

28) Click a position to the **left** of the Bracket. The Bushing is displayed.

Insert a Concentric Mate between the outside cylindrical face of the Bushing and the cylindrical face of the top right hole.

29) Click the **outside cylindrical face** of the Bushing.

30) Hold the **Ctrl** key down.

31) Click **cylindrical face** of the top right hole as illustrated.

32) Release the **Ctrl** key. The Mate Pop-up menu is displayed.

33) Click **Concentric** ⊚ from The Mate Pop-up menu. A Concentric mate locates the selected items so they can share the same center point.

Insert a Coincident Mate between the flat circular face of the Bushing and the back face of the Bracket.

34) Rotate the model to view the back face of the Bracket.

35) Click and drag the Bushing away from the back face of the Bracket.

36) If needed, Rotate the model to view the flat circular face of the Bushing.

37) Click the **flat circular face** of the Bushing.

38) Rotate the model to view the flat circular face of the Bracket.

39) Hold the **Ctrl** key down.

40) Click the **flat face** of the Bracket.

41) Release the **Ctrl** key. The Mate Pop-up menu is displayed.

42) Click **Coincident** ⟨ from The Mate Pop-up menu. A Coincident mate locates the selected faces, edges or planes so they use the same infinite lines.

Display a Trimetric view.

43) Click **Trimetric view** ▢ from the Heads-up toolbar.

Expand the Mates folder.

44) Expand the Mates folder as illustrated. View the results.

Save the model.

45) Click **Save** 💾. Note, Tangent edges are displayed on the model. To remove Tangent Edges, click Options, System Options, Display, Removed.

💡 If you delete a Mate and then recreate it, the Mate numbers will be different (increase).

💡 Determine the static and dynamic behavior of mates in each sub-assembly before creating the top-level assembly.

Activity: Insert and Mate the Third Component - Axle

Insert the Axle part into the Fly Wheel Assembly.

46) Click **Insert Component** from the Assembly toolbar. The Insert Component PropertyManager and the Window Open dialog box is displayed.

47) Double-click the **Axle** part from the SOLIDWORKS 2020\FLY WHEEL folder.

48) Click a position to the **right** of the Bracket as illustrated. The Axle is displayed.

Insert a Concentric Mate between the circular face of the Axle and the circular face of the Bushing hole.

49) **Zoom-in** on the Axle and Bushing in the Graphics window.

50) Click the **cylindrical face of the Axle**.

51) Hold the **Ctrl** key down.

52) Click the inside **cylindrical face** of the Bushing hole.

53) Release the **Ctrl** key. The Mate Pop-up menu is displayed.

54) Click **Concentric** ⊚ from the Mate Pop-up menu. Concentric2 is created.

Create a Distance Mate between the back face of the Bushing and the front face of the Axle.

55) **Rotate** the model to view the back face of the Bracket.

56) Click the **back face** of the Bushing.

57) Hold the **Ctrl** key down.

58) Click the **flat face** of the Axle as illustrated.

59) Release the **Ctrl** key. The Mate pop-up menu is displayed.

60) Click **Distance** ⊢⊣ mate from the Mate pop-up menu.

61) Enter **20**mm for Distance.

62) Click the **Green Check mark** ✔ from the Mate pop-up menu. Distance1 is created. A Distance mate locates the selected items with a specified distance between them.

Display a Trimetric view.

63) Click **Trimetric view** 🔲 from the Heads-up View toolbar.

Save the model.

64) Click **Save** 🖫 .

Activity: Insert and Mate the Fourth Component - Wheel

Insert the Wheel part into the Fly Wheel Assembly. The Wheel part was created in Chapter 2. If you did not create the part, insert the Wheel part from the SOLIDWORKS 2020\FLY WHEEL folder.

65) Click **Insert Component** from the Assembly toolbar. The Insert Component PropertyManager and Window Open dialog box is displayed.

66) Double-click the **Wheel** part from the SOLIDWORKS 2020\FLY WHEEL folder.

67) Click a **position** as illustrated in the Graphics window.

Insert a Concentric Mate between the inside cylindrical face of the hub and the outside cylindrical face of the Axle.

68) **Zoom-in** on the Wheel and Axle in the Graphics window.

69) Click the **inside cylindrical face** of the Wheel hole (Hub).

70) Hold the **Ctrl** key down.

71) Click the **cylindrical face** of the Axle.

72) Release the **Ctrl** key. The Mate Pop-up menu is displayed.

73) Click **Concentric** ⓞ from the Mate Pop-up menu.

Insert a Distance Mate between the back face of the hub and the front face of the Bushing.

74) **Rotate** the assembly to view the back face of the hub.

75) Click the **back face** of the hub as illustrated.

76) **Rotate** and **Zoom in** on the front face of the Bushing.

77) Hold the **Ctrl** key down.

78) Click the **front face** of the Bushing.

79) Release the **Ctrl** key. The Mate Pop-up menu is displayed.

80) Click **Distance** ⊢⊣ mate from the Mate pop-up menu.

81) Enter **.1**mm for Distance.

82) Click the **Green Check mark** ✔ from the Mate pop-up menu. Distance2 is created. A Distance mate locates the selected items with a specified distance between them.

Fit the model to the Graphics window.
83) Press the **f** key.

Display a Trimetric view.

84) Click **Trimetric view** ⬡ from the Heads-up View toolbar.

Save the model.
85) Click **Save** 💾 .

Rotate the Wheel in the Assembly.
86) Click on the **Wheel** and drag to view the Wheel rotate.

View the Created Mates.
87) **Expand** the Mates folder. View the created mates.

Activity: Insert and Mate the Fifth Component - Collar

Insert the Collar part into the Fly Wheel Assembly.

88) Click **Insert Component** from the Assembly toolbar. The Insert Component PropertyManager and the Window Open dialog box is displayed.

89) Double-click the **Collar** part from the SOLIDWORKS 2020\FLY WHEEL folder.

90) Click a position to the **right** of the assembly as illustrated.

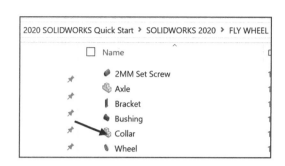

Insert a Concentric Mate between the inside cylindrical face of the Collar and the outside cylindrical face of the Axle.

91) **Zoom-in** on the Fly Wheel assembly in the Graphics window.

92) Click the **inside cylindrical face** of the Collar.

93) Hold the **Ctrl** key down.

94) Click the **outside cylindrical face** of the Axle.

95) Release the **Ctrl** key. The Mate Pop-up menu is displayed.

96) Click **Concentric** from the Mate Pop-up menu.

Insert Coincident Mate between the back face of the Collar and the front face of the hub.

97) **Rotate** the assembly to view the back face of the Collar.

98) Click the **back face** of the Collar as illustrated.

99) **Rotate** and **Zoom in** on the front face of the hub.

100) Hold the **Ctrl** key down.

101) Click the **front face** of the hub as illustrated.

102) Release the **Ctrl** key. The Mate Pop-up menu is displayed.

103) Click **Coincident** ⟨ from the Mate Pop-up menu.

Fit the model to the Graphics window.
104) Press the **f** key.

Display a Trimetric view.

105) Click **Trimetric view** 🎲 from the Heads-up View toolbar.

Rotate the Collar in the Assembly.
106) Click on the **Collar** and drag. The Collar is free to rotate.

107) Position the **2MM hole** towards the top as illustrated.

View the Created Mates.
108) **Expand** the Mates folder. View the created mates.

Activity: Insert and Mate the Sixth Component - 2MM Set Screw

Insert the 2MM Set Screw into the Collar.

109) Click **Insert Component** from the Assembly toolbar. The Insert Component PropertyManager and the Window Open dialog box is displayed.

110) Double-click the **2MM Set Screw** part from SOLIDWORKS 2020\FLY WHEEL folder.

111) Click a **position** as illustrated. The 2MM Set Screw is displayed.

Insert a Concentric Mate between the inside cylindrical face of the Collar and the outside face of the 2MM Set Screw using the Anti-Aligned option.

112) **Zoom-in** on the Collar and the 2MM Set Screw part.

113) Click the **Mate** tool from the Assembly toolbar. The Mate PropertyManager is displayed.

114) Pin the Mate PropertyManager. Click the **Keep Visible** icon.

115) Click the **inside cylindrical face** of the Collar.

116) Click the **outside cylindrical face** of the **2MM Set Screw** part.

117) If needed, click **Flip Mate Alignment (Anti-Aligned)**. Concentric mate is selected by default.

118) Click **OK** from the Concentric Mate PropertyManager.

Insert a Tangent Mate between the flat face of the 2MM Set Screw and the cylindrical face of the Axle.

119) **Rotate** and **Zoom-in** on the flat face of the 2MM Set Screw part as illustrated.

120) Drag the **2MM Set Screw** to view the bottom flat face.

121) Click the **flat face** of the 2MM Set Screw as illustrated.

Boss-Extrude1 of 2MM Set Screw<1>

122) Click the **cylindrical face** of the Axle. Tangent ○ mate is selected by default.

123) Click **OK** ✔ from the Tangent Mate PropertyManager.

124) Un-Pin the Mate PropertyManager. Click the **Keep Visible** icon as illustrated.

125) Click **OK** ✔ from the Mate PropertyManager. View the created mates in the Mates folder.

Fit the Assembly to the Graphics window.
126) Press the **f** key.

Display a Trimetric view.
127) Click **Trimetric view** 📦 from the Heads-up View toolbar.

Save the model.
128) Click **Save** 💾.

129) Click **Rebuild** and **save**. View the results.

Rotate the Axle. The Collar moves separately about the Axle. The Wheel moves separately about the Axle. Align the Wheel and Collar so they rotate with respect to the Axle. Insert a Coincident mate between the three planes (Top Plane) for proper assembly movement.

Insert a Coincident mate using the multiple mate mode between the Top Plane of the Axle, Wheel, and Collar.

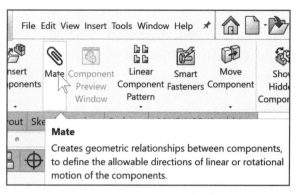

130) Click the **Mate** tool from the Assembly toolbar. The Mate PropertyManager is displayed.

131) Expand the fly-out FeatureManager from the Graphics window.

132) Expand the Axle component from the fly-out FeatureManager.

133) Click **Top Plane** from the fly-out FeatureManager.

134) Click **Multiple mate mode** from the Mate Selections box. The Components Reference box is displayed.

135) Expand the Wheel component from the fly-out FeatureManager.

136) Click **Top Plane** from the fly-out FeatureManager.

137) Expand the Collar component from the fly-out FeatureManager.

138) Click **Top Plane** from the fly-out FeatureManager. Coincident mate is selected by default.

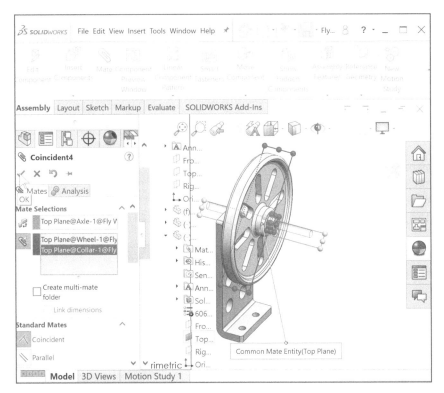

139) Click **OK** ✔ from the Coincident Mate PropertyManager.

140) **Rotate** the Wheel and view the results. Rotate the Wheel back to the original orientation where the **2MM Set Screw** part is on top.

Display a Trimetric view.

141) Click **Trimetric view** 🔲 from the Heads-up View toolbar.

Save the model.

142) Click **Save** 💾 .

Exploded View

An Exploded view shows an assembly's components spread out but positioned to show how they fit together when assembled. You create exploded views by selecting and dragging parts in the graphics area, and creating one or more explode steps. In exploded views you can:

• Evenly space exploded stacks of components (hardware, washers and so on).

• Attach a new component to the existing explode steps of another component. This is useful if you add a new part to an assembly that already has an exploded view.

• If a subassembly has an exploded view, reuse that view in a higher-level assembly.

• Add explode lines to indicate component relationships.

In the next section, create an Exploded view of the Fly Wheel assembly.

Activity: Create an Exploded View of the Fly Wheel Assembly

Create an Exploded View.
143) Click the **Assembly** tab from the CommandManager.

144) Click the **Exploded View** icon. The Explode PropertyManager is displayed. View your options.

145) Click the **Regular step (translate and rotate)** button.

Create Explode Chain1 with the 2MM Set Screw.
146) Zoom-in on the 2MM Set Screw in the Graphics window.

147) Click the **2MM Set Screw**. A Manipulator icon is displayed.

148) Zoom-out to view the assembly.

149) Drag the **Manipulator** icon upward - approximately 30mm using the Graphics window ruler.

150) Release the **Manipulator** icon.

151) Click **Done**. Chain1 is created.

Create Chain2. Select the Collar and 2MM Set Screw.
152) Click the **Collar**. A Manipulator icon is displayed.

153) Hold the **Ctrl** key down.

154) Click the **2MM Set Screw**.

155) Release the **Ctrl** key.

156) Zoom-out to view the Fly Wheel assembly.

157) Drag the **Manipulator** icon outward as illustrated - approximately 140mm using the Graphics window ruler. Click reverse direction if needed.

158) Release the **Manipulator** icon.

159) Click **Done**. Chain2 is created.

Create Chain3.

160) Click the **Wheel** from the Graphics window. A Manipulator icon is displayed.

161) Drag the **Manipulator** icon to the right of the assembly - approximately 120mm using the Graphics window ruler.

162) Release the **Manipulator** icon.

163) Click **Done**. Chain3 is created.

Create Chain4.

164) Click the **Axle** from the Graphics window. A Manipulator icon is displayed.

165) Drag the **Manipulator** icon to the right of the assembly - approximately 40mm using the Graphics window ruler.

166) Release the **Manipulator** icon.

167) Click **Done**. Chain4 is created.

Create Chain5.

168) Click the **Bushing** from the Graphics window. A Manipulator icon is displayed.

169) Drag the **Manipulator** icon to the left of the assembly - approximately 40mm - using the Graphics window ruler.

170) Click **Done**. Chain5 is created.

171) Click **OK** ✔ from the Explode PropertyManager.

View the Animate collapse of the Exploded Fly Wheel Assembly.

172) **Expand** the Default folder in the ConfigurationManager as illustrated.

173) Right-click **Exploded View1**.

174) Click **Animate collapse**. The Animation Controller is displayed.

175) Click **Playback Mode: Loop**. View the results in the Graphics window.

176) Click **Pause** .

177) Click **End** .

178) **Close** the Animation Controller.

Return to the Assembly FeatureManager.

179) Click the **FeatureManager Design Tree** icon.

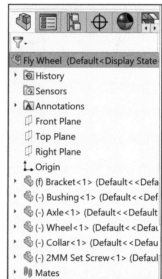

Close all documents.

180) Click **Window**, **Close All** from the Main menu. Are you finished with the assembly? The Fly Wheel assembly is a sub-assembly to the final Stirling Engine assembly.

In the next section, create the final Stirling Engine using the Fly Wheel assembly along with other provided components from the SOLIDWORKS 2020\STIRLING ENGINE folder. Use the components and sub-assemblies in the provided folder.

CommandManager and FeatureManager tabs and folder files will vary depending on system setup and Add-ins.

Activity: Create the Final Stirling Engine Assembly

Open the Sterling Engine Assembly.

181) Click **Open** from the Main menu.

182) Browse to the **SOLIDWORKS 2020\STIRLING ENGINE** folder.

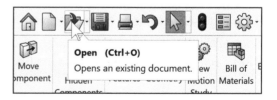

183) Double-click the **Stirling Engine** assembly. The Stirling Engine is displayed in the Graphics window.

184) **Review** the Feature Manager. The Feature Manager is made up of components. Components are sub-assemblies and parts. There are two sub-assemblies: Main Cylinder and Power. Both are displayed in a flexible state.

The Plate and Bracket are designed for flexibility to support other components in the future.

Activity: Hide the Plate Component

Hide the Plate component.

185) Right-Click **Plate** from the Assembly FeatureManager.

186) Click **Hide Components** from the Context toolbar.

Activity: Insert and Rotate the Fly Wheel Assembly

Insert the Fly Wheel assembly.

187) Click **Insert Component** from the Assembly toolbar. The Insert Component PropertyManager and the Window Open dialog box is displayed.

188) Double-click the **Fly Wheel** assembly from the SOLIDWORKS 2020\STIRLING ENGINE folder.

189) Click a **position** above the assembly. The Fly Wheel is free to translate. Drag the Fly Wheel behind the Stirling Engine assembly.

Rotate the Fly Wheel assembly.

190) Click **Rotate Component** from the Move Component drop-down menu. The Rotate Component PropertyManager is displayed.

191) **Rotate** the Fly Wheel assembly as illustrated.

192) Click **OK** ✔ from the Rotate Component PropertyManager.

Activity: Mate the Fly Wheel Assembly

Insert a Concentric mate between the cylindrical face of the Axle and
the cylindrical face of the Crank hole.

193) Click the **cylindrical face** of the Axle.

194) Click **Front view** .

195) **Rotate** and **Zoom in** on the Crank hole.

196) Hold the **Ctrl** key down.

197) Click the **inside cylindrical face** of the Crank hole as
illustrated.

198) Release the **Ctrl** key. The Mate Pop-up menu is displayed.

199) Click **Concentric** ⓞ from The Mate Pop-up menu. A
Concentric mate locates the selected items so they can
share the same center point.

Insert a Concentric mate between the cylindrical face of the Bracket hole and the cylindrical face of the Base Plate hole directly aligned above with the Axle.

200) Rotate and **Zoom** in as illustrated. If needed move the Fly Wheel assembly.

201) Click the inside **cylindrical face** of Bracket hole as illustrated.

202) Hold the **Ctrl** key down.

203) Click the **cylindrical face** of the Base Plate hole as illustrated.

204) Release the **Ctrl** key. The Mate Pop-up menu is displayed.

205) Click **Concentric** ◎ from The Mate Pop-up menu.

The Fly Wheel Bracket and the front Bracket are aligned. No additional spacers are needed to locate the Fly Wheel Bracket to the Base Plate.

View the length of the Axle.

206) Double-click on the **Axle** in the Graphics window. The Axle length is 80mm. If needed, rotate the model to view the dimension. The Axle is too short and does not go through the Crank. The Axle ends at the back face of the Bushing.

207) Click **inside** the Graphics window.

In the next section, apply the Measure tool to locate the needed length of the Axle.

Apply the Measure tool.

208) Press the **f** key to fit the model to the Graphics window.

209) Click the **Evaluate** tab from the CommandManager.

210) Click the **Measure** tool. The Measure dialog box is displayed.

211) **Zoom in** and **Rotate** on the Bushing and Crank components as illustrated.

212) Click the **back face** of the Bushing.

213) Click the **front face** of the Crank. The Normal Distance between the two faces is 13mm.

214) **Close** the Measure dialog box.

Fit the model to the Graphics window.

215) Press the **f** key.

Modify the Axle length.

216) Double click on the **Axle** in the Graphics window. View the dimensions.

217) Open the **Axle** part in the Assembly.

218) Double-click on **Boss-Extrude1**.

219) Modify the overall length from **80** to **93**.

220) **Save** the Axle.

221) **Return** to the assembly. The Axle is the correct length.

Display a Trimetric view.

222) Click **Trimetric view** from the Heads-up View toolbar.

Save the model.

223) Click **Save**.

224) Click **Save All**.

Click and drag the Handle to rotate the Crank. The Crank rotates. The linkage controls the linear motion of the Power Piston. The Main Cylinder Piston translates. The Fly Wheel is rigid and does not translate or rotate. The Fly Wheel is a sub-assembly and uses mates to rotate the Axle, Shaft Collar and Wheel. When a component is inserted into an assembly, any mates are solved as rigid.

In the next section, make the Fly Wheel flexible within the main assembly.

Make the Fly Wheel sub-assembly Flexible.

225) Right-click **Fly Wheel** from the Assembly FeatureManager.

226) Click **Make Subassembly Flexible** from the Pop-up Content toolbar.

227) Rotate the Wheel in the Graphics window. The Axle of the Fly Wheel assembly is free to rotate. The Crank is free to rotate.

In the next section, apply mates to control the rotation of the Crank and Axle.

Insert a Coincident Mate between the Top Plane of the Axle and the Top Plane of the Crank.

228) Expand the Fly Wheel component from the FeatureManager.

229) Expand the Axle component from the FeatureManager.

230) Click **Top Plane**.

231) Hold the **Ctrl** key down.

232) Expand the Crank component from the fly-out FeatureManager.

233) Click **Top Plane** from the fly-out FeatureManager.

234) Release the **Ctrl** key. The Mate Pop-up menu is displayed.

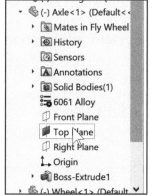

235) Click **Coincident** ⚲ mate from the Mate pop-up menu. Note if these two planes were not coincident, a parallel plane could be used to control rotation.

236) Rotate the Crank. The Fly Wheel, Crank and Axle all rotate together.

Display the Plate.

237) Right-click **Plate** from the Assembly FeatureManager.

238) Click **Show Components** for the Content toolbar. View the final assembly in the Graphics window.

Display a Trimetric view.

239) Click **Trimetric view** 🔲 from the Heads-up View toolbar.

Save the model.

240) Click **Save** 🖫.

241) Click **Rebuilt and Save**.

Pack and Go

The Pack and Go tool gathers all related files for a model design (parts, assemblies, drawings, references, design tables, Design Binder content, decals, appearances, scenes and SOLIDWORKS Simulation results) into a single folder or zip file. It's one of the best tools to utilize when you are trying to save a large assembly or drawing with references and SOLIDWORKS Toolbox components.

In the next section apply the Pack and Go tool to save the assembly and all toolbox components. The Pack and Go dialog box lists related files to be saved into a folder or zip file.

Activity: Pack and Go the Stirling Engine Assembly

Save the Assembly using Pack and Go to a Zip file.

242) Click **File**, **Pack and Go** from the Main menu bar. The Pack and Go dialog box is displayed. The dialog box lists related files to be saved into a folder or zip file. For additional information see SOLIDWORKS Help.

243) **View** your options.

244) Check **Save to Zip file** as illustrated.

245) **Browse** to the SOLIDWORKS 2020 folder.

246) Enter the **File name**.

247) Click **Save** from the Save As dialog box.

248) Click **Save** from the Pack and Go dialog box. A Zip file is created of the assembly and all reference components.

Close all models.

249) Click **Window, Close All** from the Main menu. You are finished with this chapter.

Summary

You established a SOLIDWORKS session and created two new assemblies with user defined document properties:

- Fly Wheel.

- Stirling Engine.

You inserted the following Standard and Quick mate types: Coincident, Concentric, Distance and Tangent.

You utilized the following assembly tools: Insert Component, Suppress, Un-suppress, Mate, Move Component, Rotate Component, Hide, Show, Flexible, Ridge and Multiple mate mode.

You created an Exploded View with animation and applied the Measure and Mass Properties tool.

In Chapter 4, address clearance, interference, static and dynamic behavior of the Stirling Engine

Modified Assembly.

Verify the behavior between the following components: Power Piston, Power Clevis, Connecting Rod, and Handle in the assembly.

Apply the following assembly tools: Move, Rotate, Collision Detection, Interference Detection, Selected Components, Edit Feature and Center of Mass.

Utilize the Assembly Visualization tool on the Stirling Engine Modify assembly and sort by component mass.

Create a new Coordinate System on the Stirling Engine Modify assembly relative to the default origin.

Run a Motion Study and save the Motion Study AVI file.

Questions

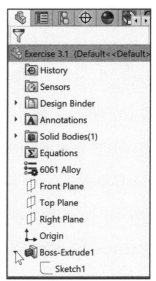

1. Identify the three default Reference planes used in SOLIDWORKS.

2. True or False. Sketches are created only on the Front Plane.

3. Identify the sketch tool required to create a hexagon.

4. Describe the profile required for an Extruded Thin feature.

5. Mirror Entities, Offset Entities, Sketch Fillet, and Trim Entities are located in the _____ toolbar.

6. List the six standard principle orthographic views in a drawing: _____, _____, _____, _____, _____, _____,

7. Identify the type of Geometric relations that can be added to a sketch.

8. Describe the difference between a Circular Pattern feature and a Linear Pattern feature.

9. Describe the difference between a Fillet feature and a Chamfer feature.

10. Identify the function of the Hole Wizard feature.

11. Four 10-24x3/8 Machine Screws are required for an assembly. The diameter is _____. The threads per inch is _____. The length is _____.

12. Describe the difference between a Distance mate and a Coincident mate.

13. True or False. A fixed component cannot move in an assembly.

14. Describe the procedure to remove the fix state (f) of a component in an assembly.

15. Determine the procedure to rotate a component in an assembly.

16. Describe the procedure to resolve a Lightweight component in an assembly.

17. Describe the procedure to resolve a Lightweight assembly.

Exercises

Exercise 3.1: HEX-NUT Part

Create an ANSI, IPS HEX-NUT part. Apply 6061 Alloy material. Apply the following dimensions:

- Depth: .125 in, [3.18].

- Inside hole diameter: .190in, [4.83].

- Outside diameter: .375in, [9.53].

Use the Top Plane as the Sketch plane.

Exercise 3.2: FRONT-SUPPORT-2 Assembly

Create an ANSI, IPS FRONT-SUPPORT-2 assembly.

- Name the new assembly FRONT-SUPPORT-2.

- Insert the FRONT-SUPPORT assembly. The FRONT-SUPPORT assembly is provided in the **Chapter 3 Homework folder**.

- Fix the FRONT-SUPPORT assembly to the Origin.

- Insert the first HEX-NUT (Exercise 3.1) into the FRONT-SUPPORT-2 assembly.

- Insert a Concentric mate and Coincident mate.

- Insert the second HEX-NUT part.

- Insert a Concentric mate and Coincident mate.

You can also insert a Parallel mate between the HEX-NUT parts and the FRONT-SUPPORT assembly.

Exercise 3.3: Weight-Hook Assembly

Create an ANSI, IPS Weight-Hook assembly. The Weight-Hook assembly has two components: WEIGHT and HOOK.

- Create a new assembly document.

- Copy and insert the WEIGHT part from the Chapter 3 Homework folder.

- Fix the WEIGHT to the Origin in the Assem1 FeatureManager.

- Insert the HOOK part from the Chapter 3 Homework folder into the assembly.

- Insert a Concentric mate between the inside top cylindrical face of the WEIGHT and the cylindrical face of the thread. Concentric is the default mate.

- Insert the first Coincident mate between the top edge of the circular hole of the WEIGHT and the top circular edge of Sweep1, above the thread.

- Coincident is the default mate. The HOOK can rotate in the WEIGHT.

- Fix the position of the HOOK. Insert the second Coincident mate between the Right Plane of the WEIGHT and the Right Plane of the HOOK. Coincident is the default mate.

- Expand the Mates folder and view the created mates. Note: Tangent edges and origins are displayed for educational purposes.

- Calculate the Mass and Volume of the assembly.

- Identify the Center of Mass for the assembly.

Exercise 3.4: Weight-Link Assembly

Create an ANSI, IPS Weight-Link assembly. The Weight-Link assembly has two components and a sub-assembly: Axle component, FLATBAR component, and the Weight-Hook sub-assembly that you created in a previous exercise.

- Create a new assembly document. Copy and insert the Axle part from the Chapter 3 Homework folder.

- Fix the Axle component to the Origin.

- Copy and insert the FLATBAR part from the Chapter 3 Homework folder. Rotate as needed.

- Insert a Concentric mate between the Axle cylindrical face and the FLATBAR inside face of the top circle.

- Insert a Coincident mate between the Front Plane of the Axle and the Front Plane of the FLATBAR.

- Insert a Coincident mate between the Right Plane of the Axle and the Top Plane of the FLATBAR. Position the FLATBAR as illustrated.

- Insert the Weight-Hook sub-assembly that you created in Exercise 3.3.

- Insert a Tangent mate between the inside bottom cylindrical face of the FLATBAR and the top circular face of the HOOK, in the Weight-Hook assembly. Tangent mate is selected by default. Click Flip Mate Alignment if needed.

- Insert a Coincident mate between the Front Plane of the FLATBAR and the Front Plane of the Weight-Hook sub-assembly. Coincident mate is selected by default. The Weight-Hook sub-assembly is free to move in the bottom circular hole of the FLATBAR.

- Calculate the Mass and Volume and the Center of Mass for the assembly.

- **Exercise 3.5: Binder Clip**

- Create a simple Gem® binder clip.

- Create an ANSI - IPS assembly.
  Create two components - Binder and
  Binder Clip.

- Apply material (plan carbon steel) to
  each component and address all
  needed mates. Think about where
  you would start.

Note: The cuts in the main body are made before folding
the metal and strengthening the material during
manufacturing. You do not need to model this exactly.
The key to this exercise is to apply Advance distance
mates (Limit mates) to limit the movement of the handle.

- What is the Base Sketch for each component?

- What are the dimensions? Measure all dimensions
  (approximately) from a small or large Gem binder
  clip.

- View SOLIDWORKS Help or Chapter 5 on the
  Swept Base feature to create the Binder Clip handle.

☼ Determine the static and dynamic behavior of mates
in each sub-assembly before creating the top level
assembly.

☼ Use the Quick mate procedure to apply Advanced distance angle
(LimitAngle) mates.

Exercise 3.6: Limit Mate (Advanced Mate)

- Copy all files from the Chapter 3 Homework\Limit Mate folder to your hard drive.

- Create the final assembly with all needed mates for proper movement.

- Insert a Distance (Advanced Limit Mate) to restrict the movement of the Slide Component - lower and upper movement.

- Use the Measure tool to obtain maximum and minimum distances.

- Use SOLIDWORKS Help for additional information.

A Distance (Limit Mate) is an Advanced Mate type. Limit mates allow components to move within a range of values for distance and angle. You specify a starting distance or angle as well as a maximum and minimum value.

- Save the model and move the slide to view the results in the Graphics window.

- Think about how you would use this mate type in other assemblies.

Use the Quick mate procedure to apply Advanced distance (Limit) mates.

Exercise 3.7: Screw Mate (Mechanical Mate)

- Copy the Chapter 3 Homework\Screw Mate folder to your hard drive.

- Open the Screw Mate assembly.

- Insert a Screw mate between the inside Face of the Base and the Face of the vice and any other mates that are required. A Screw is a Mechanical Mate type. View the avi file for proper movement.

- Calculate the total mass and volume of the assembly.

A Screw mate constrains two components to be concentric and also adds a pitch relationship between the rotation of one component and the translation of the other. Translation of one component along the axis causes rotation of the other component according to the pitch relationship. Likewise, rotation of one component causes translation of the other component. Use SOLIDWORKS Help if needed.

- Rotate the handle and view the results. Think about how you would use this mate type in other assemblies.

Use the Select Other tool (See SOLIDWORKS Help) to select faces and edges that are hidden in an assembly.

Exercise 3.8: Angle Mate

- Copy the Chapter 3 Homework\Angle Mate folder to your hard drive.

- Open the Angle Mate assembly.

- Move the Handle in the assembly. The Handle is free to rotate.

- Set the angle of the Handle.

- Insert an Angle mate (165 degree) between the Handle and the Side of the valve using Planes. An Angle mate places the selected items at the specified angle to each other. The Handle has a 165 degree Angle mate to restrict flow through the valve.

Think about how you would use this mate type in other assemblies.

Exercise 3.8A: Angle Mate (Cont:)

- Create two end caps (lids) for the ball valve using the Top-down Assembly method. Note: The Reference - In-Content symbols in the FeatureManager.

- Modify the Appearance of the body to observe the change and enhance visualization.

- Apply the Select-other tool to obtain access to hidden faces and edges.

Exercise 3.9: Symmetric Mate (Advanced Mate)

- Copy the Chapter 3 Homework\Symmetric Mate folder to your hard drive.

- Open the Symmetric Mate assembly.

- Insert a Symmetric Mate between the two Guild Rollers.

- Apply LimitMates to limit movement of the two Guild Rollers.

- Think about how you would use this mate type in other assemblies.

Exercise 3.10: Gear Mate (Mechanical mate)

View the ppt presentation and avi file on Gears located in the
Chapter 3 Homework\Gears folder.

Create the Gear Mate assembly as illustrated.

The Gear assembly consists of a **Base Plate**, two **Shafts** for the
gears, and two different gears from the **SOLIDWORKS
toolbox**.

Use the SOLIDWORKS toolbox to insert the needed gear
components. You create the **Base Plate**, and **Shafts**.

Insert all needed mates for proper movement.

Gear mate: Forces two components to rotate relative to one
another around selected axes. The Gear mate provides the ability
to establish gear type relations between components without
making the components physically mesh.

SOLIDWORKS provides the ability to modify the gear ratio
without changing the size of the gears. Align the components
before adding the Mechanical gear mate.

Exercise 3.10A: Slot Mate (Center in Slot option)

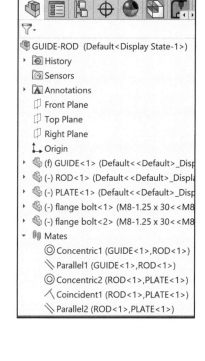

- Copy the Chapter 3 Homework\Slot Mate folder to your hard drive.

- Open the GUIDE-ROD assembly. Create two Slot mates for the two flange bolts. You can mate bolts to straight or arced slots and you can mate slots to slots. Use the Quick Mate procedure.

- Insert a Slot mate (Center in Slot option) between the first flange bolt and the inside face of the right slot.

- Insert the second slot mate (Center in Slot option) between the second flange bolt and the inside face of the left slot.

- Insert a Coincident mate between the top face of the GUIDE and the bottom face of the second flange bolt cap.

- Insert a Coincident mate between the top face of the GUIDE and the bottom face of the first flange bolt cap.

- Insert a parallel mate between the front face of the first hex head flange bolt and the front of the GUIDE.

- Inset a parallel mate between the front face of the second hex head flange bolt and the of the GUIDE.

- Expand the Mates folder. View the created mates.

Exercise 3.11: Slider Part

Create the part from the illustrated A-ANSI - MMGS Third Angle Projection drawing below: Front, Top, Right and Isometric view.

Note: The location of the Origin (shown in an Isometric view).

- Apply Cast Alloy steel for material.

- The part is symmetric about the Front Plane.

- Apply Mid Plane for End Condition in Boss-Extrude1.

- Apply Through All for End Condition in Cut-Extrude1.

- Apply Through All for End Condition in Cut-Extrude2.

- Apply Up to Surface for End Condition in Boss Extrude2

- Calculate the Volume of the part and locate the Center of mass.

Think about the steps that you would take to build the model. Do you need the Right view for manufacturing? Does it add any important information?

Exercise 3.12: Cosmetic Thread Part

- **Apply a Cosmetic thread:** ANSI 1/4-20-1 UNC 2A. A cosmetic thread represents the inner diameter of a thread on a boss or the outer diameter of a thread.

- Copy the Cosmetic thread part from the Chapter 3 Homework folder to your hard drive.

Boss-Extrude1

- Open the part.

- Create a Cosmetic thread. Produce the geometry of the thread.

- Click the bottom edge of the part as illustrated.

- Click Insert, Annotations, Cosmetic Thread from the Menu bar menu. View the Cosmetic Thread PropertyManager. Edge<1> is displayed.

- Select ANSI Inch for Standard.

- Select Machine Threads for Type.

- Select ¼-20 for Size.

- Select Blind for End Condition.

- Enter 1 for depth.

- Select 2A for Thread class.

- Click OK ✔ from the Cosmetic Thread FeatureManager.

- Expand the FeatureManager. View the Cosmetic Thread feature. If needed, right-click the Annotations folder, click Details.

- Check the Cosmetic threads and Shaded cosmetic threads box.

- Click OK. View the cosmetic thread on the model.

Exercise 3.13: Hole - Block

- Create the Hole-Block ANSI - IPS part using the Hole Wizard feature as illustrated. Create the Hole-Block part on the Front Plane.

- The Hole Wizard feature creates either a 2D or 3D sketch for the placement of the hole in the FeatureManager.

- You can consecutively place multiple holes of the same type. The Hole Wizard creates 2D sketches for holes unless you select a non-planar face or click the 3D Sketch button in the Hole Position PropertyManager.

Hole Wizard creates two sketches: a sketch of the revolved cut profile of the selected hole type and the other sketch of the center placement for the profile. Both sketches should be fully defined.

- Create a rectangular prism 2 inches wide by 5 inches long by 2 inches high. On the top surface of the prism, place four holes, 1 inch apart.

- Hole #1: Simple Hole Type: Fractional Drill Size, 7/16 diameter, End Condition: Blind, 0.75 inch deep.

- Hole #2: Counterbore hole Type: for 3/8 inch diameter Hex bolt, End Condition: Through All.

- Hole #3: Countersink hole Type: for 3/8 inch diameter Flat head screw, 1.5 inch deep.

- Hole #4: Tapped hole Type, Size ¼-20, End Condition: Blind -1.0 inch deep.

Exercise 3.14: Hole Wizard Part

- Apply the 3D sketch placement method as illustrated in the FeatureManager. Insert and dimension a hole on a cylindrical face.

- Copy the Hole Wizard 3-14 part from the Chapter 3 Homework folder to your hard drive.

- Open the Hole Wizard 3-14 part.

Note: With a 3D sketch, press the Tab key to move between planes.

- Click the Hole Wizard 📷 Features tool. The Hole Specification PropertyManager is displayed.

- Select the Counterbore Hole Type.

- Select ANSI Inch for Standard.

- Select Socket Head Cap Screw for fastener Type.

- Select 1/4 for Size. Select Normal for Fit. Select Through All for End Condition.

- Enter .100 for Head clearance in the Options box. Click the Positions Tab. The Hole Position PropertyManager is displayed.

- Click the 3D Sketch button. SOLIDWORKS displays a 3D

  interface with the Point ✏️ᵡᵞ ⟨ tool active.

🔆 When the Point tool is active, wherever you click you will create a point.

- Click the cylindrical face of the model as illustrated. The selected face is displayed in blue.

- Insert a dimension between the top face and the Sketch point.

- Click the Smart Dimension ⌁ Sketch tool.

- Click the top flat face of the model and the sketch point.

- Enter .25in.

- Locate the point angularly around the cylinder. Apply construction geometry.

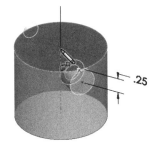

- Activate the Temporary Axes. Click View, Hide/show, check the Temporary Axes box from the Menu bar toolbar.

- Click the Line ✐ Sketch tool. Note: 3D sketch is still activated.

- Ctrl+click the top flat face of the model. This moves the red space handle origin to the selected face. This also constrains any new sketch entities to the top flat face. Note the mouse pointer

  ✐ icon.

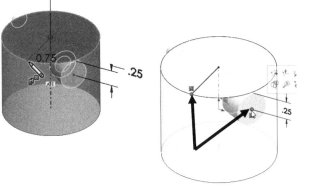

- Move the mouse pointer near the center of the activated top flat face as illustrated. View the small black circle. The circle indicates that the end point of the line will pick up a Coincident relation.

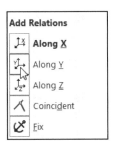

- Click the center point of the circle.

- Sketch a line so it picks up the **AlongZ sketch relation**. The cursor displays the relation to be applied. **This is a very important step. If needed, create an AlongZ sketch relation**.

Add Relations

↕ˣ	**Along X**
↕	Along Y
↕ᶻ	Along Z
⋏	Coincident
⚿	Fix

- Create an AlongY sketch relation between the centerpoint of the hole on the cylindrical face and the endpoint of the sketched line as illustrated. The sketch is fully defined.

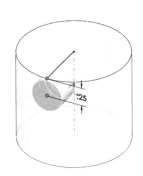

- Click OK ✔ from the Properties PropertyManager.

- Click OK ✔ from the Hole Position PropertyManager.

- Expand the FeatureManager and view the results. The two sketches are fully defined. One sketch is the hole profile; the other sketch is to define the position of the feature.

- Close the model.

Exercise 3.15: Counter Weight Assembly

Create the Counter Weight assembly as illustrated using SmartMates and Standard mates.

Copy the Chapter 3 Homework\Counter-Weight folder to your local hard drive.

Create a new assembly. The Counter Weight assembly consists of the following items:

 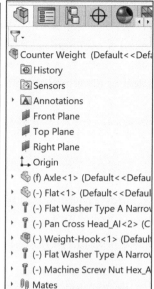

- Weight-Hook sub-assembly.

- Weight.

- Eye Hook.

- Axle component (f). Fixed to the origin.

- Flat component.

- Flat Washer Type A (from the SOLIDWORKS Toolbox).

- Pan Cross Head Screw (from the SOLIDWORKS Toolbox).

- Flat Washer Type A (from the SOLIDWORKS toolbox).

- Machine Screw Nut Hex (from the SOLIDWORKS Toolbox).

Apply SmartMates with the Flat Washer Type A Narrow_AI, Machine Screw Nut Hex_AI and the Pan Cross Head_AI components.

Use a Distance mate to fit the Axle in the middle of the Flat. Note a Symmetric mate could replace the Distance mate. Think about the design of the assembly. Apply all needed Lock mates.

The symbol (f) represents a fixed component. A fixed component cannot move and is locked to the assembly Origin.

Exercise 3.16: Gear Mate

Insert a Mechanical Gear mate between two gears. Gear mates force two components to rotate relative to one another about selected axes. Valid selections for the axis of rotation for gear mates include cylindrical and conical faces, axes, and linear edges.

- Copy the Chapter 3 Homework folder to your hard drive.

- Open Mechanical Gear Mate from the Chapter 3 Homework\Gear Mate folder.

- Click and drag the right gear.

- Click and drag the left gear. The gears move independently.

- Click the Mate ✎ tool from the Assembly tab. The Mate PropertyManager is displayed.

- Clear all selections.

- Align the components before adding the Mechanical mate. To align the two components, select a Front view and move them manually.

- Select the inside cylindrical face of the left hole.

- Select the inside cylindrical face of the right hole.

- Expand the Mechanical Mates box from the Mate PropertyManager.

- Click Gear 🕸 mate from the Mechanical Mates box. The GearMate1 PropertyManager is displayed. Accept the default settings.

- Click OK ✔ from the GearMate1 PropertyManager.

- Expand the Mates folder. View the created GearMate1.

- Rotate the Gears in the Graphics window.

- View the results.

Exercise 3.17: Cam Mate

Create a Tangent Cam follower mate and a Coincident Cam follower mate. A cam-follower mate is a type of Tangent or Coincident mate. It allows you to mate a cylinder, plane or point to a series of tangent extruded faces, such as you would find on a cam.

Create the profile of the cam from lines, arcs and splines as long as they are tangent and form a closed loop.

- Copy the Chapter 3 Homework folder to your hard drive.

- Open Cam Mate from the from the Chapter 3 Homework\Cam Mate folder. View the model. The model contains two cams.

- Expand the Mates folder to view the existing mates.

Insert a tangent Cam follower mate.

- Click the Mate ✎ tool from the Assembly tab.

- Pin the Mate PropertyManager.

- Expand the Mechanical Mates box from the Mate PropertyManager.

- Click Cam Follower ⌀ mate.

- Click inside of the Cam Path box.

- Click the outside face of cam-s<1> as illustrated.

- Click the face of Link-c<1> as illustrated. Face<2>@Link-c-1 is displayed.

- Click OK ✔ from the CamMateTangent1 PropertyManager.

Insert a Coincident Cam follower mate.

- Click Cam Follower ⟁ mate from the Mechanical Mates box.

- Click the outside face of cam-s2<1>.

- Click the vertex of riser<1> as illustrated. Vertex<1>@riser-1 is displayed.

- Click OK ✔ from the CamMateCoincident1 PropertyManager.

- Un-pin the Mate PropertyManager.

- Click OK ✔ from the Mate PropertyManager.

- Expand the Mates folder. View the new created mates.

- Display an Isometric view.

- Rotate the cams.

- View the results in the Graphics window.

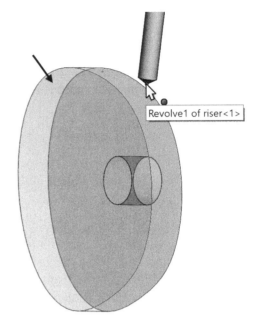

When creating a cam follower mate, make sure that your Spline or Extruded Boss/Base feature, which you use to form the cam contact face, does nothing but form the face.

In the next exercise of the spur gear and rack, the gear's pitch circle must be tangent to the rack's pitch line, which is the construction line in the middle of the tooth cut.

Exercise 3.18: Rack Pinion Mate

The Spur gear and rack components for this exercise were obtained from the SOLIDWORKS Toolbox folder marked Power Transmission. Insert a Mechanical Rack Pinion mate.

- Copy the Chapter 3 Homework folder to your hard drive.

- Open Rack Pinion Mate from the Chapter 3 Homework\Rack Pinion Mate folder.

- View the displayed: (TooCutSkeSim and ToothCutSim). If needed, display the illustrated sketches to mate the assembly.

The Diametral Pitch, Pressure Angle and Face Width values should be the same for the spur gear as the rack.

💡 Tangency is required between the spur gear and rack.

- Expand the Mates folder to view the existing mates. Four Coincident mates and a Distance mate were created in the assembly.

Insert a Rack Pinion mate.

- Click the Mate ✎ tool from the Assembly tab.

- Expand the Mechanical Mates box from the Mate PropertyManager.

- Click Rack Pinion ✖ mate from the Mechanical Mates box.

- Mate the rack. Make sure the teeth of the spur gear and rack are meshing properly, and they are not interfering with each other. It is important to create a starting point where the gear and rack have no interference.

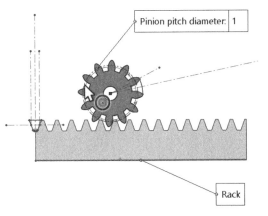

- Click the bottom linear edge of the rack in the direction of travel. Any linear edge that runs in the direction of travel is acceptable.

- Click the spur gear pitch circle as illustrated. The Pinion pitch diameter, 1in is taken from the sketch geometry. Accept the default settings.

- Click OK ✔ from the RackPinionMate1 PropertyManager.

- Rotate the spur gear on the rack. View the results.

- Hide the sketches in the model.

- Close the model.

The Pitch Diameter is the diameter of an imaginary pitch circle on which a gear tooth is designed. Pitch circles of the two spur gears are required to be tangent for the gears to move and work properly.

The distance between the center point of the gear and the pitch line of the rack represents the theoretical distance required.

🔆 Diametral Pitch is the ratio equal to the number of teeth on a gear per inch of pitch diameter.

🔆 Before applying the Rack Pinion mate, make sure the teeth of the spur gear and rack are meshing properly.

Pressure Angle is the angle of direction of pressure between contacting teeth. Pressure angle determines the size of the base circle and the shape of the involute teeth. It is common for the pressure angle to be 20 or 14 1/2 degrees.

Exercise 3.19: Hinge Mate

Create a Mechanical Hinge mate. The two components in this tutorial were obtained from the SOLIDWORKS What's New section.

- Copy the Chapter 3 Homework folder to your hard drive.

- Open Hinge Mate from the Chapter 3 Homework\Hinge Mate folder. Two components are displayed.

A Hinge mate has the same effect as adding a Concentric mate plus a Coincident mate. You can also limit the angular movement between the two components.

- Click the Mate ✎ tool from the Assembly tab.

- Expand the Mechanical Mates box from the Mate PropertyManager.

- Click Hinge ⊞ mate from the Mechanical Mates box.

Set two Concentric faces.
- Click the inside cylindrical face of the first component as illustrated.

- Click the outside cylindrical face of the second component as illustrated. The two selected faces are displayed.

Set two Coincident faces.
- Click the front flat face of the first component as illustrated.

- Click the bottom flat face of the second component as illustrated. View the results in the Graphics window.

You can specify a limit angle for rotation by checking the **Specify angle limits** box and selecting the required faces.

- Click OK ✔ from the Hinge1 PropertyManager.

- Rotate the flap component about the pin.

- Close the model.

A Hinge mate limits the movement between two components to one rotational degree of freedom.

Similar to other mate types, Hinge mates do not prevent interference or collisions between components. To prevent interference, use Collision Detection or Interference Detection.

Exercise 3.20: Collision Detection

Apply the Collision Detection tool to an assembly.

1. **Copy** the Chapter 3 Homework folder to your hard drive.

2. Open **Collision Detection** from the Chapter 3 Homework\Collision Detection folder.

3. Click the **Move Component** tool from the Assembly tab in the CommandManager. View the Move Component PropertyManager.

4. Click the **Collision Detection** box.

5. Click the **These components box**. The These components option provides the ability to select individual components for collision detection.

6. Check the **Stop at collision** box.

7. Click the **bottom face** of BASE1-1 as illustrated.

8. Click the **top face** of COLLAR111-2.

9. Click the **Resume Drag** button.

10. Click and drag the **Collar downward** towards the base as illustrated. The Collision Detection tool informs the user when there is a collision between the selected components.

11. Click **OK** from the Move Component PropertyManager.

12. **Close** the model.

Chapter 4

Design Modifications

Below are the desired outcomes and usage competencies based on the completion of Chapter 4.

Desired Outcomes:	Usage Competencies:
• Address clearance, interference, static and dynamic behavior of the Stirling Engine Modified Assembly.	• Ability to utilize the following assembly tools: Move, Rotate, Collision Detection, Interference Detection, Selected Components, Edit Feature, Center of Mass, and Assembly Visualization.
• Verify behavior between the following components: Power Piston, Power Clevis, Connecting Rod, and Handle.	• Comprehend the Interference Detection tool to locate the interference between components in an assembly.
• A new Coordinate System.	• Knowledge to locate and edit the correct mates.
• Run and save a Motion Study AVI file.	• Apply the Measure and Mass Properties tool.
	• Create a new Coordinate System relative to the origin of the Assembly.

Notes:

Chapter 4 - Design Modifications

Chapter Objective

Before you machine or create a rapid prototype of a part or assembly, you should verify the model for clearance, interference, static and dynamic behavior. Verify behavior between the Power Piston, Power Clevis, Connecting Rod and Handle with Design modifications using the Stirling Engine Modified assembly.

On the completion of this chapter, you will be able to:

- Establish a SOLIDWORKS session.

- Utilize the following assembly tools: Move, Rotate, Collision Detection, Interference Detection, Selected Components, Edit Feature, Center of Mass, Assembly Visualization and Motion Study.

- Utilize the Interference Detection tool to locate the interference between components in an assembly.

- Locate and edit the correct mate.

- Insert a Center of Mass point into the Assembly.

- Modify the Center of Mass point relative to a new Coordinate System.

- Apply the Measure and Mass Properties tool.

- Utilize the Assembly Visualization tool on the Stirling Engine Modified Assembly and sort components by mass.

- Run a Motion Study.

- Create and save the Motion Study AVI file.

Activity: Start a SOLIDWORKS Session

Start a SOLIDWORKS session.
1) Double-click the **SOLIDWORKS icon** from the desktop.

2) If needed, **close** the SOLIDWORKS Welcome dialog box.

In the next section, open the Stirling Engine Modified assembly from the SOLIDWORKS 2020\Design Modifications folder. Verify for clearance, interference, static and dynamic behavior between the Power Piston, Power Clevis, Connecting Rod and Handle with Design modifications using the Stirling Engine Modified assembly.

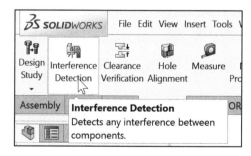

Activity: Open the Stirling Engine Modified Assembly

Open the Stirling Engine Modified Assembly.

3) Click **File**, **Open** from the Main menu.

4) **Browse** to the SOLIDWORKS 2020\Design Modifications folder.

5) Double-click the **Stirling Engine Modified** assembly. The assembly is displayed in the Graphics window.

Verify for clearance, interference, static and dynamic behavior between the Power Piston, Power Clevis, Connecting Rod and Handle in the Stirling Engine Modified Assembly.

You can detect collisions with other components when moving or rotating a component. The software can detect collisions with the entire assembly or a selected group of components.

You can find collisions for either the selected components or for all of the components that move as a result of mates to the selected components.

Physical Dynamics is an option in Collision Detection that allows you to see the motion of assembly components in a realistic way.

With Physical Dynamics enabled, when you drag a component, the component applies a force to components that it touches and moves the components if they are free to translate or rotate.

Activity: Verify Collision between Components in the Assembly

Activate the Collision Detection tool.

6) Click the **Assembly** tab from the CommandManager.

7) Click the **Move Component** icon. The Move Component PropertyManager is displayed.

8) Click the **Collision Detection** button.

9) Check the **Highlight faces, Sound** and **Ignore complex surfaces** box from the Advanced Options section.

10) Uncheck the **Stop at collision** box.

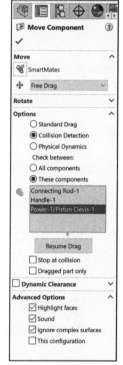

11) Click the **These components** button. If needed Right-click inside the Components for Collision Check box and click **Clear Selections**.

12) Click the **Handle** from the Graphics window. Handle-1 is displayed in the Components for collision Check box.

13) Click the **Connecting Rod** from the Graphics window. Connection Rod-1 is displayed in the Components for Collision Check box.

14) Click the **Piston Clevis** as illustrated from the Graphics window. Power-1/Piston Clevis 1 is displayed in the Components for collision Check box.

15) Click the **Power Piston** as illustrated from the Graphics window. Power-1/Power Piston-1 is displayed in the Components for Collision Check box.

16) Click the **Resume Drag** button.

17) Drag the **Handle collar (Boss-Extrude1 of Collar <1>)** downward and stop when the components collide. Collision components are displayed in the Graphics window. There are collisions between the Connecting Rod and the Power Piston and the Connecting Rod and the Piston Clevis.

18) Click **OK** ✔ from the Move Component PropertyManager.

In the next section, utilize the Interference Detection tool to locate the interference between components.

Activity: Utilize the Interference Detection tool

Activate the Interference Detection tool.

19) Click the **Evaluate** tab from the CommandManager.

20) Click the **Interference Detection** icon. The Interference Detection PropertyManager is displayed.

21) Right-click in the **Selected Components** box.

22) Click **Clear Selections**.

Select component to perform the Interference detection process.

23) Click the **Connecting Rod** from the Graphics window as illustrated. Connection Rod-1 is displayed in the Selected Components box.

24) **Rotate** the model to view the Piston Clevis.

25) Click the **Piston Clevis** from the Graphics window as illustrated.

26) Click the **Power Piston** from the Graphics window as illustrated.

27) Click the **Calculate** button. There are volume interferences between the components indicated in red. *The volume interference value and the number of interferences may be different depending on the position of the Crank.*

28) Click each **Interference Results** to view the interference between the components displayed in red.

29) Click **OK** ✅ from the Interference Detection PropertyManager.

The Interference Detection and Collision Detection results require design changes to the Power Piston, Piston Clevis, Connecting Rod and other components in the Stirling Engine Modified Assembly. Review Assembly and Evaluate tools to modify the design intent.

Activity: Modify the Connecting Distance2 Rod Mate

View the Mates on the Connecting Rod in the Assembly.

30) **Expand** Connecting Rod in the FeatureManager.

31) **Expand** the Mates in Stirling Engine Modified folder as illustrated.

32) Click on each **Mate** to review how the Connecting Rod is constrained in the Stirling Engine Modified assembly.

Modify the Distance Mate between the face of the Rod and the face of the Connection Rod.

33) Right-click **Distance2** from the FeatureManager.

34) Click the **Edit Feature** 🔧 icon from the Context toolbar. The current distance is 0mm.

35) Enter **3mm** for distance.

36) Click **OK** ✔ from the Distance2 PropertyManager.

Display a Trimetric view.

37) Click **Trimetric view** 🔲 from the Heads-up toolbar.

Save the model.

38) Click **Save** 💾.

39) Click **Rebuild and Save the document**.

40) Click **Save All**.

💡 A Distance Mate of 0mm allows flexibility when you need to vary spacing between components.

💡 If you delete a Mate and then recreate it, the Mate numbers will be different (increase).

Utilize the Measure tool to check the Distance mate in the Graphics window.

41) Click the **Evaluate** tab in the CommandManager.

42) Click the **Measure** tool. The Measure dialog box is displayed.

43) Click the **Show XYZ Measurements** button.

44) Right-click in the **Selection** box.

45) Click **Clear Selections**.

46) **Zoom in** on the two faces (Rod and Connecting Rod).

47) Click the **Rod face**. Click the **Connecting Rod face**. The Normal Distance is 3.00mm.

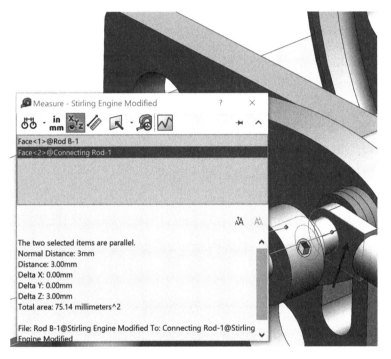

Close the Measure Dialog box.

48) Click **Close** ☒.

Display a Trimetric view.

49) Click **Trimetric view** from the Heads-up toolbar.

Save the model.

50) Click **Save** .

Activity: Utilize the Interference Detection tool - Check the Solution

Activate the Interference Detection tool.

51) Click the **Evaluate** tab from the CommandManager.

52) Click the **Interference Detection** icon. The Interference Detection PropertyManager is displayed.

53) Right-click in the **Selected Components** box.

54) Click **Clear Selections**.

Select component to perform the Interference detection process.

55) Click the **Connecting Rod** from the Graphics window as illustrated. Connection Rod-1 is displayed in the Selected Components box.

56) **Rotate** the model to view the Piston Clevis.

57) Click the **Piston Clevis** from the Graphics window as illustrated.

58) Click the **Power Piston** from the Graphics window as illustrated.

59) Click the **Calculate** button. There are no interferences between the components. The modification of the Distance mate resolved the interference issue.

60) Click **OK** ✓ from the Interference Detection PropertyManager.

Activity: Locate the Center of Mass of the Assembly

Locate the Center of Mass along with other Mass Properties of the assembly (COM). You can assign values for mass, center of mass and moments of inertia to override the calculated values.

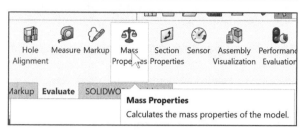

View the Mass Properties of the Assembly.
61) Click the **Evaluate** tab from the CommandManager.

62) Position the **Handle** vertically upward in the assembly.

63) Click the **Mass Properties** icon. The Mass Properties dialog box is displayed. View the results. The total mass is 3447.14 grams. The numbers represent the document properties (2 decimal places). Your Center of mass may be different due to the location of the Handle in the assembly.

View the Mass Properties of the Plate in the Assembly.
64) Right-click inside the **Selection** box.

65) Click **Clear Selections**.

66) Click the **Plate** in the Graphics window as illustrated. Plate-1 is displayed in the Selection box. The Mass is approximately 259.60 grams.

Return to the Mass Properties of the Assembly.
67) Right-click inside the **Selection** box.

68) Click **Clear Selections**.

69) Click **Stirling Engine Modified** from the Assembly FeatureManager as illustrated.

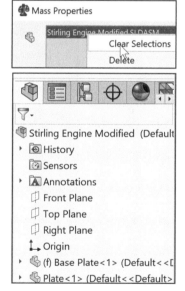

Locate the Center of Mass of the Assembly.
70) Click the **Create Center of Mass feature** box.

Close the Mass Properties dialog box.
71) Click **Close** ⊠.

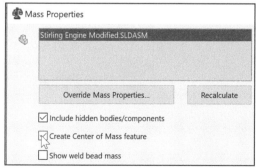

Display the COM point in the Assembly.

72) If needed, click **View Center of Mass** from the Hide/Show Items in the Heads-up toolbar.

73) Click a **position** in the Graphics window. View the results.

Display a Trimetric - Shaded With Edges view.

74) Press the **space bar** to display the Orientation dialog box.

75) Click **Trimetric** . You can also access the Isometric view tool from the Heads-up View toolbar.

76) Click the **Shaded With Edges** icon.

Save the Assembly.

77) Click **Save** .

78) Click **Rebuild** and **save**.

You can add a Center of Mass (COM) point to parts and assemblies. In drawings of parts or assemblies that contain a COM point, you can show and reference the COM point.

You can define a coordinate system for a part or assembly. Coordinate systems are useful:

- With the Measure and Mass Properties tools.

- When exporting SOLIDWORKS documents to IGES, STL, ACIS, STEP, Parasolid, VRML and VDA.

- When applying assembly mates.

In the next section, create a new coordinate system on the front bottom vertex of the assembly.

Activity: Create a new Coordinate System in the Assembly

View the new Coordinate System on the front bottom vertex of the assembly.

79) Click the **Assembly** tab from the CommandManager.

80) Click **Coordinate System** ⊹ from the Consolidated Reference Geometry drop-down menu. The Coordinate System PropertyManager is displayed.

81) Click the **bottom front vertex** of the Assembly Base Plate. Vertex<1> is displayed.

82) Click **OK** ✔ from the Coordinate System PropertyManager. Coordinate System1 is displayed in the Assembly FeatureManager.

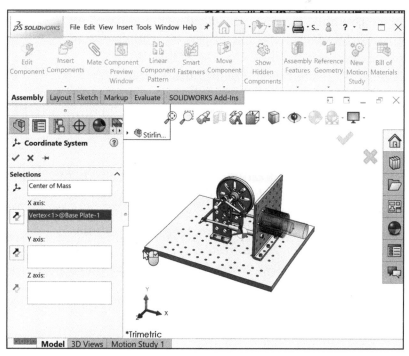

In the next section, display the Mass Properties relative to the new Coordinate system of the assembly.

Activity: Display the Mass Properties relative to the new Coordinate System

View the new Center of Mass point.

83) Click the **Evaluate** tab from the CommandManager.

84) Click the **Mass Properties** ⚖ icon. The Mass Properties dialog box is displayed.

85) Click the **drop-down arrow** for the Report coordinate values relative to:

86) Select **Coordinate System1**. View the updated results (Center of Mass) in the Mass Properties dialog box. Your Center of mass may be different due to the original location of the Handle in the assembly.

Close the Mass Properties dialog box.

87) Click **Close** ☒.

Display a Trimetric view.

88) Press the **space bar** to display the Orientation dialog box.

89) Click **Trimetric view** ⬦. You can also access the Isometric view tool from the Heads-up View toolbar.

Assembly Visualization

In an assembly, the designer selects material based on cost, performance, physical properties, manufacturing processes, sustainability, etc. The SOLIDWORKS Assembly Visualization tool includes a set of predefined columns to help troubleshoot assembly performance. You can view the open and rebuild times for the components, and the total number of graphics triangles for all instances of components.

The Assembly Visualization tool provides the ability to rank components based on the default values (**weight, mass, density, volume, etc.**) or their custom properties (**cost, sustainability, density, surface area, volume, etc.**) or an equation and activate a spectrum of colors that reflects the relative values of the properties for each component.

Hide/Show Value Bar icon. Available for numeric properties. Turns the value bars off and on. When the value bars are on, the component with the highest value displays the longest bar. You can set the length of the bars to be calculated relative to the highest-value component or relative to the entire assembly.

Flat Nested View icon. Nested view, where subassemblies are indented. Flat view, where subassembly structures are ignored (similar to a parts-only BOM).

Grouped/Ungrouped View icon. Groups multiple instances of a component into a single line item in the list. Grouped View is useful when listing values for properties that are identical for every instance of the component. Ungrouped view lists each instance of a component individually. Ungrouped View is useful when listing values for instance-specific properties, such as Fully mated, which might be different for different instances of the component.

Performance Analysis icon. Provides additional information on the open, display, and rebuild performance of models in an assembly.

Filter icon. Filters the list by text and by component show/hide state.

The Assembly Visualization tab in the FeatureManager design tree panel contains a list of all components in the assembly, sorted initially by file name. There are three default columns:

- File name.
- Quantity.
- Mass.

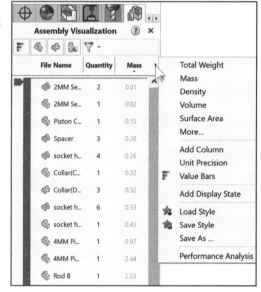

Mass is the default column. View your options as illustrated other than Mass.

You can save the list information in a separate file such as a Microsoft® Excel® spreadsheet or a text file.

In the next section, utilize the Assembly Visualization tool on the Stirling Engine Modify Assembly. Sort the assembly components by mass.

Activity: Assembly Visualization tool - Sort Components by Mass

Apply the Assembly Visualization tool. Sort by mass.

90) Click the **Evaluate** tab from the CommandManager.

91) Click the **Assembly Visualization** icon. The Assembly Visualization PropertyManager is displayed.

92) Click the **arrow** to the right of the File Name row as illustrated.

93) Select **Mass**.

94) Click the **Mass** column.

95) **View** the results from low to high or from high to low.

96) Click the **Toggle Slider** to view the color of the components in the Graphics window.

97) Explore the other **columns** and **options**.

98) **View** the results in the Graphics window.

Return to the Assembly FeatureManager.

99) Click the Assembly

FeatureManager Design tree icon.

Display a Trimetric view.

100) Press the **space bar** to display the Orientation dialog box.

101) Click **Trimetric view** . You can also access the Isometric view tool from the Heads-up View toolbar.

Save the Assembly.

102) Click **Save** .

Motion Study

Motion studies are graphical simulations of motion for assembly models. You can incorporate visual properties such as lighting and camera perspective into a motion study.

Motion studies do not change an assembly model or its properties. They simulate and animate the motion you prescribe for a model. You can use SOLIDWORKS mates to restrict the motion of components in an assembly when you model motion.

In the next section, play the Basic Motion study of the final assembly. You can use Basic Motion for approximating the effects of motors, springs, contact and gravity on assemblies. Basic Motion takes mass into account in calculating motion. Basic Motion computation is relatively fast, so you can use this for creating presentation-worthy animations using physics-based simulations. A simple rotary motor was applied to the Handle for movement.

> ### Activity: Play the Basic Motion Study of the Assembly

Play the Basic Motion Study. Access the Motion Manager from the Motion Study tab.

103) Click the **Motion Study #** tab at the bottom left of the Graphics window. The Motion Manager is displayed.

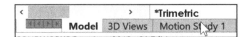

104) Select **Basic Motion** from the Motion Study drop-down menu. A motion study was created using a simple Rotary Motor on the Handle.

105) Click the **Calculate** icon.

106) Click the **Play from Start** ▮►. View the Motion Study in the Graphics window.

Create an AVI file of the Motion Study.

107) Click the **Save Animation** icon. View the Save Animation to File dialog box.

108) Select a **Save in** location. View your options.

109) Click **Save**. View your options.

110) Click **OK** from the Video Compression dialog box.

Return to the Model.

111) Click the **Model** tab at the bottom left of the Graphics window.

112) Explore SOLIDWORKS Help for additional information on Basic Motion using timers and other tools.

Save the Assembly.

113) Click **Save** 💾.

114) Click **Rebuild** and **save**.

Close all models.

115) Click **Window**, **Close All** from the Main menu bar.

You are finished with this section.

Summary

In this chapter, you addressed clearance, interference, static and dynamic behavior of the Stirling Engine Modified assembly.

You verified the behavior between the following components: Power Piston, Power Clevis, Connecting Rod and Handle in the assembly.

You applied the following assembly tools: Move, Rotate, Collision Detection, Interference Detection, Selected Components, Edit Feature and Center of Mass along with utilizing the Assembly Visualization tool on the Stirling Engine assembly and sorted by component mass.

You created a new Coordinate System on the Stirling Engine assembly relative to the default origin, ran a Motion Study and saved the Motion Study AVI file.

In Chapter 5, you will learn about the
Drawing and Dimension Fundamentals
and create two new drawings with user
defined document properties and custom
properties:

- Fly Wheel Assembly.

- Bushing.

Create the Fly Wheel Assembly drawing
with an Exploded Isometric view.

Utilize a Bill of Materials and Balloons.

Learn about Custom Properties and the
Title Block.

Create the Bushing Part drawing utilizing
Third Angle Projection with two standard
Orthographic views: Front, Top and an
Isometric view.

Address imported dimensions from the
Model Items tool.

Insert additional dimensions using the
Smart Dimension tool along with all
needed annotations.

Questions

1. Describe the differences between a Distance Mate and a Coincident Mate. Provide an example of both.

2. Describe an assembly or sub-assembly. Are they the same?

3. Explain how to determine an interference between components in an assembly. Provide an example.

4. Describe Mates. Why are Mates important in assembling components?

5. In an assembly, each component has_____# degrees of freedom. Name them.

6. Why would you use a Distance Mate of 0.

7. True or False. If you delete a Mate and then recreate it, the Mate numbers will be different (increase).

8. Describe the procedure in locating the Center of Mass for a part or assembly.

9. Describe the procedure in creating a new coordinate system.

10. Explain the Assembly Visualization tool. How would this tool be useful in a weight reduction process.

Exercises

Exercise 4.1: Counter Weight Assembly

- Copy the Chapter 4 Homework\Counter Weight folder to your hard drive.

- Create the Counter Weight assembly as illustrated using SmartMates and Standard mates from the Assembly FeatureManager. All components are supplied in the Chapter 8 Homework\Counter-Weight folder.

- Weight-Hook sub-assembly.

- Weight.

- Eye Hook.

- Axle component (f). Fixed to the origin.

- Flat component.

- Flat Washer Type A (from the SOLIDWORKS Toolbox).

- Pan Cross Head Screw (from the SOLIDWORKS Toolbox).

- Flat Washer Type A (from the SOLIDWORKS toolbox).

- Machine Screw Nut Hex (from the SOLIDWORKS Toolbox).

Apply SmartMates with the Flat Washer Type A Narrow_AI, Machine Screw Nut Hex_AI and the Pan Cross Head_AI components.

Use a Distance mate to fit the Axle in the middle of the Flat. Note a Symmetric mate could replace the Distance mate.

Think about the design of the assembly. Apply all needed Lock mates. The symbol (f) represents a fixed component. A fixed component cannot move and is locked to the assembly Origin.

Additional Project Exercises

Exercise 4.2: Butterfly Valve Assembly Project

Use the web to locate a similar assembly and to understand the parts needed in the assembly. View the provided sample parts. Create your own parts and final assembly. View the provided avi file to create the proper movement in SOLIDWORKS.

Create an ANSI Butterfly valve assembly document. Insert all needed components and mates to assemble the assembly. Create and insert any additional components if needed.

Create a A-ANSI Landscape - Third Angle Isometric Exploded Drawing document with Explode lines of the assembly using your knowledge of SOLIDWORKS. Insert a BOM with Balloons. Insert all needed General notes and Custom Properties in the Title Block.

Arbor Press Project Sample Part examples
Bench Vice Assembly Project Part examples
Butterfly Valve Assembly Project Part examples
Drill Guide Assembly Project Part examples
Kant Twist Clamp Assembly Project Part examples
Pipe Vice Assembly Project Part examples
Pulley Assembly Project Part examples
Quick Acting Clamp Assembly Project Part examples
Radial Engine Assembly Project Part examples
Shaper Tool Head Assembly Project Part examples
Shock Assembly Project Part examples
Welder Arm Assembly Project Part examples

ITEM NO.	PartNo	MATERIAL	DESCRIPTION	QTY.
1	1001	Gray Cast Iron	BODY	1
2	1002	Stainless Steel (ferritic)	SHAFT	1
3	1005	1060 Alloy	PLATE 2	1
4	1008	1060 Alloy	BOLT [small]	2
5	1003	1060 Alloy	PLATE	1
6	1004	1060 Alloy	BOLT	3
7	1007	Stainless Steel (ferritic)	ARM	1
8	1009	Stainless Steel (ferritic)	HANDLE	1
9	1010	1060 Alloy	HEX NUT	1

TITLE:

FINAL ASSEMBLY

SIZE A   DWG. NO. 2001   REV A

School Name

SCALE: 1:10 WEIGHT: SHEET 1 OF 3

Exercise 4.3: Shock Assembly Project

Use the web to locate a similar assembly and to understand the parts needed in the assembly. View the provided sample parts. Create your own parts and final assembly. View the provided avi file to create the proper movement in SOLIDWORKS.

Create an ANSI Shock assembly document. Insert all needed components and mates to assemble the assembly. Create and insert any additional components if needed.

Create a C-ANSI Landscape - Third Angle Isometric Exploded Drawing document with Explode lines of the assembly using your knowledge of SOLIDWORKS. Insert a BOM with Balloons. Insert all needed General notes and Custom Properties in the Title Block.

- Arbor Press Project Sample Part examples
- Bench Vice Assembly Project Part examples
- Butterfly Valve Assembly Project Part examples
- Drill Guide Assembly Project Part examples
- Kant Twist Clamp Assembly Project Part examples
- Pipe Vice Assembly Project Part examples
- Pulley Assembly Project Part examples
- Quick Acting Clamp Assembly Project Part examples
- Radial Engine Assembly Project Part examples
- Shaper Tool Head Assembly Project Part examples
- Shock Assembly Project Part examples
- Welder Arm Assembly Project Part examples

ITEM NO.	PART NUMBER	DISCRIPTION	material	QTY.
1	MPIgr3	BRACKET	ALLOY STEEL	1
2	MPIgr4	SPACER	ALLOY STEEL	1
3	MPIgr6	USUPPORT	ALLOY STEEL	1
4	MPIgr5	SHAFT	ALLOY STEEL	1
5	MPIgr2	FLAT WASHER .625-I UNC	ALLOY STEEL	1
6	MPIgr1	SELF-LOC UNC NUT	ALLOY STEEL	1
7	MPIgr9	PIVOT	ALLOY STEEL	1
8	MPIgr8	BUSHING	Commercial bronze UNS C22000 (PD-ID bronze)	2
9	MPIgr7	HEX HEAD BOLT	ALLOY STEEL	1
10	MPIgr10	CASTLE NUT	ALLOY STEEL	1
13	MPIgr11	COTTER PIN	ALLOY STEEL	1
14	MPIgr12	SPRING	AISI 304	1

.25

.25

YOUR SCHOOL NAME

SHOCK ASSEMBLY

C   CSize   A

Exercise 4.4: Quick Acting Clamp Assembly Project

Use the web to locate a similar assembly and to understand the parts needed in the assembly. View the provided sample parts. Create your own parts and final assembly. View the provided avi file to create the proper movement in SOLIDWORKS.

Create an ANSI Quick Acting clamp assembly document. Insert all needed components and mates to assemble the assembly. Create and insert any additional components if needed.

Create a C-ANSI Landscape - Third Angle Isometric Exploded Drawing document with Explode lines of the assembly using your knowledge of SOLIDWORKS. Insert a BOM with Balloons. Insert all needed General notes and Custom Properties in the Title Block.

Arbor Press Project Sample Part examples
Bench Vice Assembly Project Part examples
Butterfly Valve Assembly Project Part examples
Drill Guide Assembly Project Part examples
Kant Twist Clamp Assembly Project Part examples
Pipe Vice Assembly Project Part examples
Pulley Assembly Project Part examples
Quick Acting Clamp Assembly Project Part examples
Radial Engine Assembly Project Part examples
Shaper Tool Head Assembly Project Part examples
Shock Assembly Project Part examples
Welder Arm Assembly Project Part examples

ITEM NO.	PART NUMBER	DESCRIPTION	Material	QTY.
1	100	LEFT SUPPORT	Alloy Steel	1
2	WPI-07200	PIN A	Alloy Steel	3
3	2133	ARM	Alloy Steel	2
4	34566	PIN B	Alloy Steel	1
5	34465	GRIP	Alloy Steel	1
6	WPI-063477	HINGE B	Alloy Steel	1
7	3eer4	SHAFT	Alloy Steel	1
8	WPI-05556y6	HINGE A	Alloy Steel	1
9	WPI-1133	RIGHT SUPPORT	Alloy Steel	1
10	WPI-03	HANDLE	Alloy Steel	1
11	WPI-04	HANDLE RING	Alloy Steel	2

QUICK ACTING CLAMP

C   334-447   A

Exercise 4.5: Drill Guide Assembly Project

Use the web to locate a similar assembly and to understand the parts needed in the assembly. View the provided sample parts. Create your own parts and final assembly. View the provided avi file to create the proper movement in SOLIDWORKS.

Create an ANSI Drill Guide assembly document. Insert all needed components and mates to assemble the assembly. Create and insert any additional components if needed.

Create a C-ANSI Landscape - Third Angle Isometric Exploded Drawing document with Explode lines of the assembly using your knowledge of SOLIDWORKS. Insert a BOM with Balloons. Insert all needed General notes and Custom Properties in the Title Block.

Arbor Press Project Sample Part examples
Bench Vice Assembly Project Part examples
Butterfly Valve Assembly Project Part examples
Drill Guide Assembly Project Part examples
Kant Twist Clamp Assembly Project Part examples
Pipe Vice Assembly Project Part examples
Pulley Assembly Project Part examples
Quick Acting Clamp Assembly Project Part examples
Radial Engine Assembly Project Part examples
Shaper Tool Head Assembly Project Part examples
Shock Assembly Project Part examples
Welder Arm Assembly Project Part examples

ITEM NO.	PART NUMBER	DESCRIPTION	MATERIAL	QTY.
1	WPI-1000-01	BASE	2014 ALLOY	1
2	WPI-1000-02	ROTATOR	PLAIN CARBON STEEL	2
3	WPI-1000-03	ROD GUIDE	CAST ALLOY STEEL	2
4	WPI-1000-04	SLIDE	2014 ALLOY	1
5	WPI-1000-05	COLLAR	2014 ALLOY	2
6	WPI-1000-06	BUSHING	ALUMINUM BRONZE	2
7	WPI-1000-07	RETAINING RING B27.1 - NA2-45	ALLOY STEEL	2
8	WPI-1000-08	DRILL ADAPTOR	PLAIN CARBON STEEL	1
9	WPI-1000-09	THUMB SCREW .25-20x0.51 TYPE B, FLAT POINT-C	2014 ALLOY	6

Exercise 4.6: Pulley Assembly Project

Use the web to locate a similar assembly and to understand the parts needed in the assembly. View the provided sample parts. Create your own parts and final assembly. View the provided avi file to create the proper movement in SOLIDWORKS.

Create an ANSI Pully assembly document. Insert all needed components and mates to assemble the assembly. Create and insert any additional components if needed.

Create a A-ANSI Landscape - Third Angle Isometric Exploded Drawing document with Explode lines of the assembly using your knowledge of SOLIDWORKS. Insert a BOM with Balloons. Insert all needed General notes and Custom Properties in the Title Block.

Note: Additional components were added below.

- Arbor Press Project Sample Part examples
- Bench Vice Assembly Project Part examples
- Butterfly Valve Assembly Project Part examples
- Drill Guide Assembly Project Part examples
- Kant Twist Clamp Assembly Project Part examples
- Pipe Vice Assembly Project Part examples
- Pulley Assembly Project Part examples
- Quick Acting Clamp Assembly Project Part examples
- Radial Engine Assembly Project Part examples
- Shaper Tool Head Assembly Project Part examples
- Shock Assembly Project Part examples
- Welder Arm Assembly Project Part examples

ITEM NO.	PART NUMBER	DESCRIPTION	MATERIAL	QTY.
1	6	Pulley Base	Balsa	1
2	5	Bracket	Alloy Steel	2
3	2	Brace	Alloy Steel	4
4	8	Hex Nut	Alloy Steel	6
5	4	Support'	Alloy Steel	2
6	9	Screw Cap	Alloy Steel	4
7	10	Turn Screw	Alloy Steel	2
8	3	Bushing	Alloy Steel	2
9	1	Pulley	Alloy Steel	2
10	7	Washer	Alloy Steel	2
11	11	Belt1-1	Silicon Rubber	1

School Name

TITLE: **Double Pulley Assembly**

SIZE	DWG. NO.	REV
A	2223-77	A

SCALE: 1:5   WEIGHT: SHEET 1 OF 2

UNLESS OTHERWISE SPECIFIED:
DIMENSIONS ARE IN MILLIMETERS
TOLERANCES:
ANGULAR: MACH± 0°30'
ONE PLACE DECIMAL ±0.5
TWO PLACE DECIMAL ±.15

INTERPRET GEOMETRIC
TOLERANCING PER: ASME Y14.5

	NAME	DATE
DRAWN	CSG	10/5/18
CHECKED		
ENG APPR.		
MFG APPR.		
Q.A.		
COMMENTS:		

MATERIAL

FINISH

NEXT ASSY   USED ON
APPLICATION   DO NOT SCALE DRAWING

PROPRIETARY AND CONFIDENTIAL
THE INFORMATION CONTAINED IN THIS DRAWING IS THE SOLE PROPERTY OF <INSERT COMPANY NAME HERE>. ANY REPRODUCTION IN PART OR AS A WHOLE WITHOUT THE WRITTEN PERMISSION OF <INSERT COMPANY NAME HERE> IS PROHIBITED.

Exercise 4.7: Welder Arm Assembly Project

Use the web to locate a similar assembly and to understand the parts needed in the assembly. View the provided sample parts. Create your own parts and final assembly. View the provided avi file to create the proper movement in SOLIDWORKS.

Create an ANSI Welder Arm assembly document. Insert all needed components and mates to assemble the assembly. Create and insert any additional components if needed.

Create a C-ANSI Landscape - Third Angle Isometric Exploded Drawing document with Explode lines of the assembly using your knowledge of SOLIDWORKS. Insert a BOM with Balloons. Insert all needed General notes and Custom Properties in the Title Block.

- Arbor Press Project Sample Part examples
- Bench Vice Assembly Project Part examples
- Butterfly Valve Assembly Project Part examples
- Drill Guide Assembly Project Part examples
- Kant Twist Clamp Assembly Project Part examples
- Pipe Vice Assembly Project Part examples
- Pulley Assembly Project Part examples
- Quick Acting Clamp Assembly Project Part examples
- Radial Engine Assembly Project Part examples
- Shaper Tool Head Assembly Project Part examples
- Shock Assembly Project Part examples
- Welder Arm Assembly Project Part examples

Exercise 4.8: Radial Engine Assembly Project

Use the web to locate a similar assembly and to understand the parts needed in the assembly. View the provided sample parts. Create your own parts and final assembly. View the provided avi file to create the proper movement in SOLIDWORKS.

Create an ANSI Radial Engine assembly document. Insert all needed components and mates to assemble the assembly. Create and insert any additional components if needed.

Create a C-ANSI Landscape - Third Angle Isometric Exploded Drawing document with Explode lines of the assembly using your knowledge of SOLIDWORKS. Insert a BOM with Balloons. Insert all needed General notes and Custom Properties in the Title Block.

Note: Additional components were added to the drawing.

Exercise 4.9: Bench Vice Assembly Project

Use the web to locate a similar assembly and to understand the parts needed in the assembly. View the provided sample parts. Create your own parts and final assembly. View the provided avi file to create the proper movement in SOLIDWORKS.

Create an ANSI Bench Vice assembly document. Insert all needed components and mates to assemble the assembly. Create and insert any additional components if needed.

Create an A-ANSI Landscape - Third Angle Isometric Exploded Drawing document with Explode lines of the assembly using your knowledge of SOLIDWORKS.
Insert a BOM with Balloons. Insert all needed General notes and Custom Properties in the Title Block.

Arbor Press Project Sample Part examples
Bench Vice Assembly Project Part examples
Butterfly Valve Assembly Project Part examples
Drill Guide Assembly Project Part examples
Kant Twist Clamp Assembly Project Part examples
Pipe Vice Assembly Project Part examples
Pulley Assembly Project Part examples
Quick Acting Clamp Assembly Project Part examples
Radial Engine Assembly Project Part examples
Shaper Tool Head Assembly Project Part examples
Shock Assembly Project Part examples
Welder Arm Assembly Project Part examples

ITEM NO.	PART NUMBER	DESCRIPTION	MATERIAL	QTY.
1	WPI-0001	Base	Alloy Steel	1
2	WPI-0002	Base Plate	Alloy Steel	2
3	WPI-0003	Vice Jaw	Alloy Steel	1
4	WPI-0004	Clamping Plate	Alloy Steel	1
5	WPI-0005	Jaw Screw	Alloy Steel	1
6	WPI-0006	Screw Bar	Alloy Steel	1
7	WPI-0007	Bar Globes	Alloy Steel	2
8	WPI-008	Hex Screw M6x1.0x14-14WN	Alloy Steel	4
9	WPI-0009	Bolt	Alloy Steell	1
10	WPI-00019	Blot1	Alloy Steel	2

Exercise 4.10: Kant Twist Clamp Assembly Project

Use the web to locate a similar assembly and to understand the parts needed in the assembly. View the provided sample parts. Create your own parts and final assembly. View the provided avi file to create the proper movement in SOLIDWORKS.

Create an ANSI Kant Twist clamp assembly document. Insert all needed components and mates to assemble the assembly. Create and insert any additional components if needed.

Create a C-ANSI Landscape - Third Angle Isometric Exploded Drawing document with Explode lines of the assembly using your knowledge of SOLIDWORKS. Insert a BOM with Balloons. Insert all needed General notes and Custom Properties in the Title Block.

Note: Additional components were added to the drawing.

Exercise 4.11: Arbor Press Assembly Project

Use the web to locate a similar assembly and to understand the parts needed in the assembly. View the provided sample parts. Create your own parts and final assembly. View the provided avi file to create the proper movement in SOLIDWORKS.

Create an ANSI Arbor Press assembly document. Insert all needed components and mates to assemble the assembly. Create and insert any additional components if needed.

Create a A-ANSI Landscape - Third Angle Isometric Exploded Drawing document with Explode lines of the assembly using your knowledge of SOLIDWORKS. Insert a BOM with Balloons. Insert all needed General notes and Custom Properties in the Title Block.

- Arbor Press Project Sample Part examples
- Bench Vice Assembly Project Part examples
- Butterfly Valve Assembly Project Part examples
- Drill Guide Assembly Project Part examples
- Kant Twist Clamp Assembly Project Part examples
- Pipe Vice Assembly Project Part examples
- Pulley Assembly Project Part examples
- Quick Acting Clamp Assembly Project Part examples
- Radial Engine Assembly Project Part examples
- Shaper Tool Head Assembly Project Part examples
- Shock Assembly Project Part examples
- Welder Arm Assembly Project Part examples

ITEM NO.	PART NUMBER	DESCRIPTION	Material	QTY.
1	1000-14	ARBOR PRESS SUBASSEMBLY		1
1.1	1000-6	COLUMN	AISI 1020	1
1.2	1000-2	FRONT COVER	AISI 1020	1
1.3	1000-7	HEX_SCREW_AI	AISI 1035 Steel (SS)	4
2	1000-1	TABLE PLATE	AISI 1035 Steel (SS)	1
3	1000-5	STUBBY PEG	AISI 1035 Steel (SS)	1
4	1000-3	SQUARE PEG	AISI 1035 Steel (SS)	1
5	1000-4	HANDLE	AISI 1035 Steel (SS)	1
6	1000-8	Part1^ARBOR PRESS	AISI 1035 Steel (SS)	1

School Name

TITLE: ARBOR PRESS

SIZE A | DWG. NO. 100-1111 | REV A

SCALE: 1:5 WEIGHT: SHEET 1 OF 1

Exercise 4.12: Shaper tool head Assembly Project

Use the web to locate a similar assembly and to understand the parts needed in the assembly. View the provided sample parts. Create your own parts and final assembly. View the provided avi file to create the proper movement in SOLIDWORKS.

Create an ANSI Shaper tool head assembly document. Insert all needed components and mates to assemble the assembly. Create and insert any additional components if needed.

Create a C-ANSI Landscape - Third Angle Isometric Exploded Drawing document with Explode lines of the assembly using your knowledge of SOLIDWORKS. Insert a BOM with Balloons. Insert all needed General notes and Custom Properties in the Title Block.

- Arbor Press Project Sample Part examples
- Bench Vice Assembly Project Part examples
- Butterfly Valve Assembly Project Part examples
- Drill Guide Assembly Project Part examples
- Kant Twist Clamp Assembly Project Part examples
- Pipe Vice Assembly Project Part examples
- Pulley Assembly Project Part examples
- Quick Acting Clamp Assembly Project Part examples
- Radial Engine Assembly Project Part examples
- Shaper Tool Head Assembly Project Part examples
- Shock Assembly Project Part examples
- Welder Arm Assembly Project Part examples

Exercise 4.13: Quick Acting Clamp Assembly Project

Use the web to locate a similar assembly and to understand the parts needed in the assembly. View the provided sample parts. Create your own parts and final assembly. View the provided avi file to create the proper movement in SOLIDWORKS.

Create an ANSI Quick acting clamp assembly document. Insert all needed components and mates to assemble the assembly. Create and insert any additional components if needed.

Create an A-ANSI Landscape - Third Angle Isometric Exploded Drawing document with Explode lines of the assembly using your knowledge of SOLIDWORKS. Insert a BOM with Balloons. Insert all needed General notes and Custom Properties in the Title Block.

- Arbor Press Project Sample Part examples
- Bench Vice Assembly Project Part examples
- Butterfly Valve Assembly Project Part examples
- Drill Guide Assembly Project Part examples
- Kant Twist Clamp Assembly Project Part examples
- Pipe Vice Assembly Project Part examples
- Pulley Assembly Project Part examples
- Quick Acting Clamp Assembly Project Part examples
- Radial Engine Assembly Project Part examples
- Shaper Tool Head Assembly Project Part examples
- Shock Assembly Project Part examples
- Welder Arm Assembly Project Part examples

ITEM NO.	PART NUMBER	DESCRIPTION	Material	QTY.
1	10	LEFT SUPPORT	Alloy Steel	1
2	07	PIN A	Alloy Steel	3
3	01	ARM	Alloy Steel	2
4	08	PIN B	Alloy Steel	1
5	I-02	GRIP	Alloy Steel	1
6	I-06	HINGE B	Alloy Steel	1
7	I-09	SHAFT	Alloy Steel	1
8	I-05	HINGE A	Alloy Steel	1
9	I-11	RIGHT SUPPORT	Alloy Steel	1
10	I-03	HANDLE	Alloy Steel	1
11	I-04	HANDLE RING	Alloy Steel	2

QUICK ACTING CLAMP

Notes:

Chapter 5

Drawing and Dimensioning Fundamentals

Below are the desired outcomes and usage competencies based on the completion of Chapter 5.

Desired Outcomes:	Usage Competencies:

• An Isometric Exploded Assembly Drawing.
 	
    • Bill of Materials.
 	
    • Balloons.
 	
• A Part Drawing.
 	
    • Front, Top, Right, & Isometric view.
 	
    • Dimensions.
 	
    • Annotations.
	• Produce a Sheet Format with Custom Sheet Properties, Title block and more.
	
	• Create an Exploded Isometric Assembly drawing view with a Bill of Materials and Balloons.
	
	• Create standard Orthographic (Front, Top, & Right) views.
	
	• Aptitude to import dimensions from the Model Items tool.
	
	• Knowledge to insert additional dimensions using the Smart Dimension toolbar.
	
	• Ability to insert drawing annotation.

Notes:

Chapter 5 - Drawing and Dimensioning Fundamentals

Chapter Objective

Create two new drawings: Fly Wheel assembly and Bushing part. The first drawing is an Isometric Exploded view of the Fly Wheel assembly. The assembly drawing displays a Bill of Materials at the part level along with Balloons, Magnetic lines, and Custom Properties.

The second drawing is a Bushing part. Insert standard Orthographic (Front, Top & Right) views along with an Isometric view. Insert all needed dimensions, annotations and Custom Properties.

On the completion of this chapter, you will be able to:

- Create a new SOLIDWORKS drawing.

- Set user defined document properties for a drawing.

- Produce a Sheet Format with Custom Sheet Properties, Title block and more.

- Generate an Exploded Isometric Assembly drawing view.

- Create and insert a Bill of Materials with Auto Balloons in the Assembly drawing.

- Produce a Part drawing with standard Orthographic (Front, Top & Right) views and an Isometric view.

- Address imported dimensions from the Model Items tool.

- Insert additional dimensions using the Smart Dimension toolbar.

- Modify drawing view display styles.

- Insert drawing view annotations.

Activity: Start a SOLIDWORKS Session

Start a SOLIDWORKS session.
1) Double-click the **SOLIDWORKS icon** from the desktop.

2) **Close** the SOLIDWORKS Welcome dialog box.

3) **Pin** ⚲ the Menu bar as illustrated. Use both the Menu bar and the Menu bar toolbar.

The foundation of a new SOLIDWORKS drawing is the Drawing Template. Drawing size, drawing standards, company information, manufacturing, and/or assembly requirements, units and other properties are defined in the Drawing Template.

Activity: Create a New Drawing Document

Create a New Drawing document.

4) Click **New** ⬜ from the Menu bar or click **File, New** from the Menu bar menu. The New SOLIDWORKS Document dialog box is displayed. Advanced mode is used in this book.

5) Double-click **Drawing** from the Templates tab. The Sheet Format/Size dialog box is displayed.

6) If needed, un-check the **Only show standard formats** box.

7) Check the **Display sheet format** box. The default Standard sheet size is A (ANSI) Landscape.

8) Click **OK** from the Sheet Format/Size dialog box.

9) Click **Cancel** ✖ from the Model View PropertyManager. Draw1 FeatureManager is displayed.

🔅 If the Start command when creating new drawing option is checked, the Model View PropertyManager is selected by default.

Draw1 is the default drawing name. Sheet1 is the default first sheet name. The default Drawing

CommandManager tabs will vary depending on system setup and Add-ins.

The Sheet Format is incorporated into the Drawing Template. The Sheet Format contains the border, Title block information, revision block information, company name, and/or company logo information, Custom Properties, and SOLIDWORKS Properties. Custom Properties and SOLIDWORKS Properties are shared values between documents.

In the next section, set Sheet Properties in the Assembly drawing document.

Activity: Set Sheet Properties for the Drawing

Set Sheet Properties for the Drawing.

10) Right-click on **Sheet1** in the Drawing FeatureManager.

11) Click **Properties** 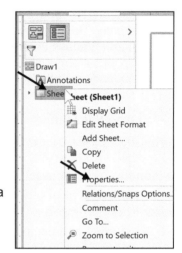. The Sheet Properties dialog box is displayed.

12) Select **A (ANSI) Landscape** from the Standard sheet size box.

13) Select **Third angle** for Type of projection.

14) Click **Apply Changes or Cancel** from the Sheet Properties dialog box. The Apply Changes command is only available if a change was made. The Sheet Properties are set for the drawing document.

There are two different types of Angle Projection: First and Third Angle Projection.

- First Angle Projection is used primarily in Europe, South American and Asia.

- Third Angle Projection is used in the United States.

Both First Angle and Third Angle projections result in the same six principle views (Front, Top, Right, Bottom, Left, and Back); the difference between them is the arrangement of these views.

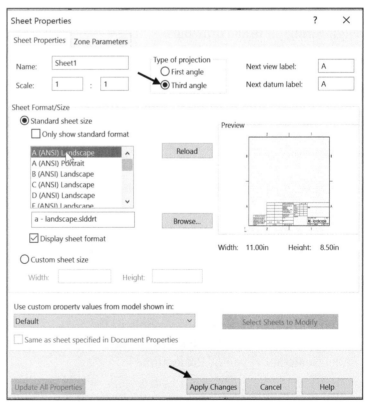

In the next section, set Document Properties in the drawing.

Activity: Set Document Properties for the Drawing

Set Document Properties. Select overall drafting standard.

15) Click **Options** ⚙ ▾ from the Menu bar. The System Options General dialog box is displayed.

16) Click the **Document Properties** tab.

17) Select **ANSI** from the Overall drafting standard drop-down menu. Various Detailing options are available depending on the selected standard.

The Overall drafting standard determines the display of dimension text, arrows, symbols, and spacing. Units are the measurement of physical quantities. Millimeter dimensioning and decimal inch dimensioning are the two most common unit types specified for engineering parts and drawings.

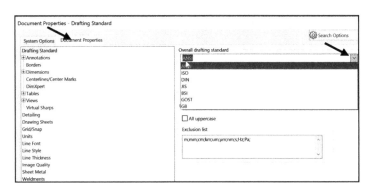

Set Document Properties. Select units and precision.

18) Click the **Units** folder.

19) Click **MMGS** (millimeter, gram, second) for Unit system.

20) Select **.12** (two decimal places) for Length basic units.

21) Click **OK** from the Document Properties - Units dialog box. The Part FeatureManager is displayed.

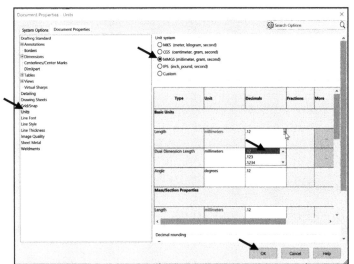

Document Properties control the display of dimensions, annotations and symbols in the drawing. Sheet Properties control Sheet Size, Sheet Scale and Type of Projection.

Title Block

The Title block contains text fields linked to System Properties and Custom Properties. System Properties are determined from the SOLIDWORKS documents. Custom Property values are assigned to named variables. Save time. Utilize System Properties and define Custom Properties in your Sheet Formats.

System Properties and Custom Properties for Title Block:			
System Properties Linked to fields in default Sheet Formats:	Custom Properties of drawings linked to fields in default Sheet Formats:		Custom Properties of parts and assemblies linked to fields in default Sheet Formats:
SW-File Name (in DWG. NO. field):	CompanyName:	EngineeringApproval:	Description (in TITLE field):
SW-Sheet Scale:	CheckedBy:	EngAppDate:	Weight:
SW-Current Sheet:	CheckedDate:	ManufacturingApproval:	Material:
SW-Total Sheets:	DrawnBy:	MfgAppDate:	Finish:
	DrawnDate:	QAApproval:	Revision:
	EngineeringApproval:	QAAppDate:	

The drawing document contains two modes:

- Edit Sheet Format.

- Edit Sheet.

Insert views and dimensions in the Edit Sheet mode. Modify the Sheet Format text, lines, and the Title block information in the Edit Sheet Format mode. The CompanyName Custom Property is located in the Title block above the TITLE box. There is no value defined for CompanyName. A small text box indicates an empty field. Define a value for the Custom Property CompanyName. Example: D&M EDUCATION LLC. The Tolerance block is located in the Title block.

A part cannot be inserted into a drawing when the Edit Sheet Format mode is selected.

The Tolerance block provides information to the manufacturer on the minimum and maximum variation for each dimension on the drawing.

If a specific tolerance or note is provided on the drawing, the specific tolerance or note will override the information in the Tolerance block. General tolerance values are based on the design requirements and the manufacturing process. Modify the Tolerance block in the Sheet Format for ASME Y14.5 machined parts. Delete unnecessary text. The FRACTIONAL text refers to inches. The BEND text refers to sheet metal parts.

Fly Wheel Assembly Drawing

A drawing contains part views, geometric dimensioning and tolerances, notes, and other related design information. When a part is modified, the drawing automatically updates. When a driving dimension in the drawing is modified, the part is automatically updated.

Drawings consist of one or more views produced from the part or assembly. Create the Fly Wheel Assembly drawing from the Fly Wheel assembly. Utilize the Task Pane to insert an Isometric Exploded drawing view. Create a Bill of Materials at the part level. Unlike a part drawing, no dimensions are required at this time. Utilize the SOLIDWORKS 2020\Drawing folder for models in the section.

Activity: Create a Fly Wheel Assembly Drawing

Insert the Isometric Exploded Drawing view of the Fly Wheel Assembly.

22) Click the **View Palette** ⬜ icon from the SOLIDWORKS Resource Pane.

23) Click the **Browse** button. The Windows Open dialog box is displayed.

24) Double-click the **Fly Wheel** assembly from the SOLIDWORKS 2020\Drawing folder.

Insert the Isometric Exploded view.

25) Drag and drop the *Isometric Exploded view icon into Sheet1. The view is displayed. Note: The Isometric Exploded view was created in Chapter 4.

Set Sheet Properties for the Drawing.

26) Right-click in the **Graphics window** (not in the Exploded Isometric drawing view).

27) Expand the drop-down menu.

28) Click **Properties** 📄. The Sheet Properties dialog box is displayed.

29) Enter Scale: **1:2**.

30) Click **Apply Changes** from the Sheet Properties dialog box. View the results in the Graphics window.

Modify the Display Style.

31) Click **inside** the Exploded Isometric drawing view. The Drawing View1 PropertyManager is displayed.

32) Click **Shaded with Edges** 🔲 for Display Style.

33) Click the **Green Check mark** ✔ from the Drawing View1 PropertyManager.

Save the Fly Wheel assembly drawing.

34) Click **Save As** 🔳.

35) Click **Yes**.

36) **Select** the save in folder.

37) **Enter** Fly Wheel for drawing name.

38) Click **Save**.

Balloons

You do not have to insert a Bill of Materials in order to add balloons. You can use auto balloons to automatically generate balloons in a drawing view.

Insert a set of balloons into one or more drawing views. Balloons are inserted into the appropriate views without duplicates. You can specify that balloons be inserted when creating new drawing views. You can specify that the balloons follow the assembly order or are numbered sequentially. You can automatically insert magnetic lines when using the Auto Balloon command.

In the next section, insert Auto Balloons using Magnetic lines in the assembly.

Activity: Insert Auto Balloons into the Assembly Drawing

Insert Auto Balloons with Magnetic lines.

39) Click inside the **Exploded Isometric drawing** view. The Drawing View 1 PropertyManager is displayed.

40) Click the **Annotation** tab from the CommandManager.

41) Click the **Auto Balloon** icon.

42) **View** the balloons in the Graphics window. Note: Your location may vary.

43) Click **OK** from the Auto Balloon PropertyManager.

44) **View** the results.

45) **Click** and **drag** the Balloons as illustrated.

46) If needed, click **OK** ✔ from the Balloon PropertyManager.

💡 Magnetic lines are a convenient way to align balloons along a line at any angle. You attach balloons to magnetic lines, choose to space the balloons equally or not and move the lines freely at any angle in the drawing.

Bill of Materials

SOLIDWORKS automatically populates a Bill of Materials (BOM) with item numbers, quantities, part numbers, and custom properties in assembly drawings if they are inserted in the proper document. You can anchor, move, edit, and split a BOM.

When you insert balloons into a drawing, the item numbers and quantities in the balloons correspond to the numbers in the Bill of Materials. If an assembly has more than one configuration, you can list quantities of components for all configurations or selected configurations.

You can create BOMs in assembly files and multi-body part files. You can insert a BOM saved with an assembly into a referenced drawing. You do not need to create a drawing first.

In the next section, insert a Bill of Materials (BOM).

Activity: Insert a Bill of Materials

Insert a Bill of Materials at the part level.

47) Click inside the **Exploded Isometric view** in the Graphics window.

48) Click the **Annotation** tab from the CommandManager.

49) Click the **Tables** drop-down menu.

50) Click the **Bill of Materials** icon. The Bill of Materials PropertyManager is displayed.

51) Click the ⭐ **Open table template for Bill of Materials** icon.

52) Double-click **bom-material.sldbomtbt**.

53) Check the **Parts only** BOM Type.

54) Click **OK** ✔ from the Bill of Materials PropertyManager.

55) Click a **position** in the upper left corner to place the Bill of Material Table.

56) **View** the results. Move the Isometric Exploded view if needed.

Save the Fly Wheel assembly drawing.

57) Click **Save** 💾.

CommandManager tabs will vary depending on system setup and Add-ins.

In the next section address Custom Properties of the parts in the Bill of Materials. Later, address Custom Properties of the Drawing and Title block.

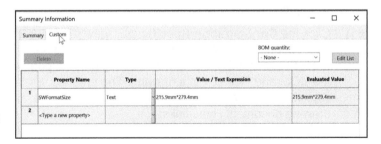

Activity: Address Custom Properties of the Parts in the Bill of Materials

Open the Axle part from the drawing.

58) Right-click the **Axle** in the drawing view.

59) Click **Open Part**. The Axle part is displayed. The Axle FeatureManager is displayed. Material is 6061 Alloy. For the Material to be displayed in a drawing or Bill of Materials, you need to set the Custom Properties for each part in an assembly along with other needed information for the drawing. Note: In this case, it's set for you.

View Custom Properties for the Axle part.

60) Click **File**, **Properties** from the Menu bar. The Summary Information dialog box is displayed.

61) Click the **Configuration Specific** tab. View the entered information: *Description* and *Material*. You would need to enter this information for all new created parts to populate the BOM and Title box in a drawing. The entered Description and Material information is related to the drawing document Hyperlinks.

Delete Row1 & 2. Re-enter Description and Material information as an exercise.

62) Click in **Row 2**. Press the **Delete** button.

63) Click in **Row1**. Press the **Delete** button.

64) Click the **drop-down arrow** under Property Name as illustrated.

65) Select **Description**.

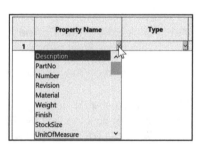

66) Click inside the **Value / Text Expression** box.

67) Enter **AXLE**.

68) Click the **drop-down arrow** under Property Name in Row 2.

69) Select **Material**.

70) Click the **drop-down arrow** under Value / Text Expression.

71) Select **Material**. View the results. The Material drawing link is displayed. 6061 Alloy is displayed in the Evaluated Value box.

72) Click **OK** from the Summary Information dialog box.

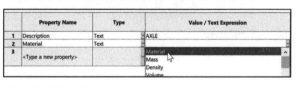

73) **Close** the AXLE part. **Return** to the drawing. Address Custom Properties for the drawing.

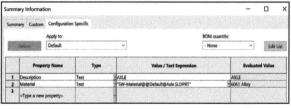

Activity: Address Custom Properties - Modify the Title block

Define Custom Properties in the drawing document.

74) Click **File**, **Properties** from the Menu bar. The Summary Information dialog box is displayed.

75) Click the **Custom** tab.

76) **Select and enter** the illustrated items. Note: Enter the present DrawnDate and your DrawnBy initials. Enter your school's name under CompanyName.

	Property Name	Type	Value / Text Expression	Evaluated Value
1	SWFormatSize	Text	215.9mm*279.4mm	215.9mm*279.4mm
2	CompanyName	Text	D&M ENGINEERING	D&M ENGINEERING
3	Revision	Text	A	A
4	DrawnBy	Text	DCP	DCP
5	DrawnDate	Text	9-1-2019	9-1-2019

77) Click **OK** from the Summary Information dialog box. View the results.

Activate the Edit Sheet Format mode.

78) **Right-click** in Sheet1. Do not click inside the Isometric drawing view.

79) Click **Edit Sheet Format**. The Title block lines turn blue.

80) **Zoom in** on the Title block.

Modify the Tolerance Note.

81) Double-click the text **INTERPRET GEOMETRIC TOLERANCING PER:**.

82) Enter **ASME Y14.5**.

83) Click **OK** ✓ from the Note PropertyManager.

84) Double-click inside the **Tolerance block** text. The Formatting dialog box and the Note PropertyManager is displayed.

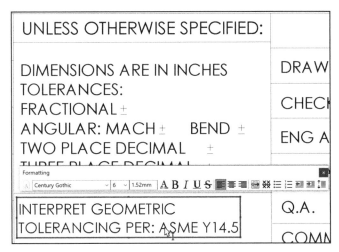

85)   Delete the text **INCHES**.

86)   Enter **MILLIMETERS**.

87)   Delete the line **FRACTIONAL ±** as illustrated.

88)   Delete the text **BEND ±** as illustrated.

89)   Click a position at the end of the **ANGULAR: MACH ±** line.

90)   Enter **0**.

91)   Click the **Add Symbol** button from the Text Format box. The Symbols dialog box is displayed.

92)   Click **Degree** from the drop-down menu.

93)   Enter **30′** for minutes of a degree.

94)   Delete the **TWO** and **THREE PLACE DECIMAL** lines.

95)   Enter **ONE PLACE DECIMAL ±0.5**.

96)   Enter **TWO PLACE DECIMAL ±0.15**.

97)   Click **OK** from the Note PropertyManager.

Insert a Company Logo and the Third Angle Projection icon.

98)   Click **Insert**, **Picture** from the Menu bar.

99)   Double-click the **LOGO.jpg** from the Drawing folder. Note, if you have your own Logo, use it for the drawing.

100)   The Sketch Picture PropertyManager is displayed. Un-check the **Lock aspect ratio** box and the **Enable scale tool** box.

101)   Drag the **picture handles** to size the picture to the left of the Title block as illustrated.

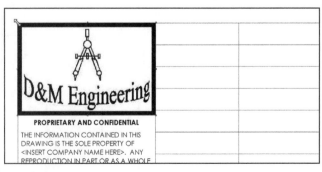

102) Click **OK** ✔ from the Sketch Picture PropertyManager.

103) Click **Insert**, **Picture** from the Menu bar. The Open dialog box is displayed.

104) Double-click the **Third Angle Projection.jpg**. The Sketch Picture PropertyManager is displayed.

105) Drag the **picture handles** to size the icon as illustrated.

106) Click **OK** ✔ from the Sketch Picture PropertyManager.

Return to the Edit Sheet mode.

107) Right-click in the Graphics window.

108) Click **Edit Sheet**. View the results.

Save the Fly Wheel assembly drawing.

109) Click **Save** 💾. As an exercise, change the DWG. NO to 555-66 as shown below. Use the drawing Custom Properties, and the Link to Properties tool.

Bushing Part Drawing

A drawing contains part views, geometric dimensioning and tolerances, notes, and other related design information. When a part is modified, the drawing automatically updates. When a driving dimension in the drawing is modified, the part is automatically updated.

Create a Bushing part drawing with inserted dimensions and annotations. Modify the Display mode to show Hidden Lines.

Drawings consist of one or more views produced from the part or assembly. Insert four drawing views: Front, Top, Right and Isometric. Hide any unnecessary drawing views. The Front view should be the most descriptive view. It should be positioned in a natural orientation based on its function.

Open and start the Bushing Drawing with an empty Bushing Drawing document located in the SOLIDWORKS 2020\Drawing folder.

Activity: Open the Empty Bushing Part Drawing

Open the Empty Bushing Part Drawing.

110) Click **File**, **Open** from the Main menu.

111) Browse to the **SOLIDWORKS 2020\Drawing** folder.

112) Double-click the **Bushing** part drawing. The empty Bushing part drawing is displayed. Third Angle Projection is selected for Projection type. ANSI is the drafting Standard. MMGS are the units.

There are two different types of Angle Projection: First and Third Angle Projection.

- First Angle Projection is used in Europe, South American and Asia.

- Third Angle Projection is used in the United States.

Both First Angle and Third Angle projections result in the same six principle views (Front, Top, Right, Bottom, Left, and Back); the difference between them is the arrangement of these views.

Insert four drawing views: Front, Top, Right and Isometric. Utilize the View Palette in the Task Pane. All drawing views are projected from the Front view. You can place any view in the Front view location. The Front view should be the most natural and descriptive view of the model.

Insert four drawing views.

113) Click the **View Palette** ⊞ icon from the SOLIDWORKS Resources Pane.

114) Click the **Browse** button.

115) Browse to the **SOLIDWORKS 2020\Drawing** folder.

116) Double-click the **Bushing** part.

117) **Drag** and **drop** the (A) Right view into the Front view location as illustrated.

118) Click directly **upwards** to create the Top view**.**

119) Click directly **downward** and to the **right** to create the Right view.

120) Click a location **approximately 45 degrees** to the upper right of the Front view. This is the Isometric view.

121) Click **OK** ✔ from the PropertyManager. View the four views.

122) Click and drag the **drawing views** to space them for dimensions and annotations. The Top and Right view are directly projected from the Front view.

Save the Bushing part drawing.

123) Click **Save** 💾.

124) Click **Save All**.

Drawing Dimensions

Dimension the part drawing views. Utilize the Model items tool to automatically import model dimensions. Utilize the Smart Dimension tool to insert any additional needed dimensions.

🔆 In SOLIDWORKS, inserted dimensions in the drawing are displayed in gray. Imported dimensions from the part are displayed in black.

Import Model dimensions using the Model Items tool.

125) Click the **Annotation** tab.

126) Click the **Model Items** ✎ tool. The Model Items PropertyManager is displayed. View your options and tool.

127) Select **Entire model** from the Source drop-down menu.

128) Click the **Hole Wizard Locations** ⊤ icon.

129) Click the **Hole callout** ⌴⌀ icon. This option inserts hole callout annotations to hole wizard features.

130) Click **OK** ✅ from the Model Items PropertyManager. Imported part dimensions are displayed in the drawing.

💡 Dimensions are imported into the drawing. The dimensions may not be in the correct location with respect to the feature or size of the part per the ANSI Y-14.5 standard. Move them later in the chapter and address extension line gaps (5mm) and annotations (Centerline).

💡 Move parent (Front) and child (Top and Right) views independently by dragging their view boundary. Hold the Shift key down and select multiple views to move as a group.

Third Angle Projection type symbol is illustrated.

The dimensions and text in the next section have been enlarged for visibility. Drawing dimension location is dependent on *Feature dimension creation* and *Selected drawing views*.

Move dimensions within the same view. Use the mouse pointer to drag dimensions and leader lines to a new location. Leader lines reference the size of the profile. A gap must exist between the profile lines and the leader lines. Shorten the leader lines to maintain a drawing standard. Use the blue Arrow buttons to flip the dimension arrows.

Notes provide relative part or assembly information. Example: Material type, material finish, special manufacturing procedure or considerations, preferred supplier, etc.

Below are a few helpful guidelines to create general drawing notes:

- Use Upper Case letters.

- Use left text justification.

- Font size should be the same as the dimension text.

Hide superfluous feature dimensions. Do not delete feature dimensions. Recall hidden dimension with the View, Show Annotations command. Move redundant, dependent views outside the sheet boundary or hide redundant views.

Full cylinders (holes and bosses) must always be measured by their diameter. The diameter symbol must precede the numerical value to indicate that the dimension shows the diameter of a circle or cylinder. The symbol used is the Greek letter phi (Ø).

 If a hole goes completely through the feature and it is not clearly shown on the drawing, the abbreviation "**THRU**" or "**THRU ALL**" in all upper case follows the dimension.

You should not dimension a hole in a drawing by its radius.

Move and Hide drawing dimensions in the Front view.
131) Zoom-in on the Front view. Right-click and hide all dimensions in the Front view.

132) Right-click **Ø5** in the Front view.

133) Click **Hide**.

134) Right-click **Ø12** in the Front view.

135) Click **Hide**.

136) Right-click **Ø10** in the Front view.

137) Click **Hide**.

138) Right-click the second **Ø5** dimension.

139) Click **Hide.**

140) Click the **Annotations** tab. Click the **Smart Dimension** tool. Insert a **Ø5** dimension in the Front view with a bent leader line as illustrated.

Insert Dimension Text.

141) Click **Ø5** in the Front View. The Dimension PropertyManager is displayed.

142) Click a **position after** <MOD-DIM><DIM> in the Dimension Text box.

143) Press the **space** key.

144) Enter **THRU**.

145) Click **OK** ✔ from the Dimension PropertyManager. View the results in Drawing view1.

Display Hidden Lines Visible.

146) Click **inside Drawing View2** as illustrated. The Drawing View2 PropertyManager is displayed. **Move** the dimensions as illustrated.

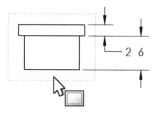

147) Click **Hidden Lines Visible** ⬡ from the Display Style box.

148) Click **OK** ✔ from the Drawing View2 PropertyManager. View the results.

💡 General practice is to stagger the dimension text on parallel dimensions (small to large).

Arrowheads are used to terminate dimension lines.

Insert needed feature and location dimensions in the Top view.

149) Zoom-in on the Top view.

150) Click the **6mm** dimension.

151) Move the 6mm dimension to the left of the drawing view as illustrated.

152) Click the **control points** on the extension lines to create 5mm gaps as illustrated.

All drawing views with dimensions require gaps (5mm) between the visible feature line and extension lines.

Insert a diameter dimension in the Top View.

153) Click **Smart Dimension** from the CommandManager.

154) Click the **top horizontal line** as illustrated.

155) Click **position** above the model. View the results.

Insert a second diameter dimension in the Top View.

156) Click the **bottom horizontal line**.

157) Click a position **below** the model. View the results.

158) Click **OK** from the Dimension PropertyManager.

Centerlines or Center marks should be used on all circles, holes and slots.

Insert a Centerline in the Top View.

159) Click **inside Drawing View2**. The Drawing View2 PropertyManager is displayed.

160) Click **Centerline** ⊞ from the Annotation CommandManager. The Centerline PropertyManager is displayed.

161) Click the **Select View** box.

162) Click **OK** ✔ from the Centerline PropertyManager. View the results. Move any dimensions if needed.

Display Shaded With Edges (Drawing View4).

163) Click **inside Drawing View4** as illustrated. The Drawing View4 PropertyManager is displayed.

164) Click **Shaded with Edges** ⬛ from the Display Style box.

165) Click **OK** ✔ from the Drawing View4 PropertyManager. View the results.

Save the Bushing part drawing.

166) Click **Save** 💾.

167) Click **Yes**.

Hide the Right View.

168) Right-click in **Drawing View3**.

169) **Expand** the drop-down menu. Click **Hide**. Hide unneeded drawing views. The Right View is not needed.

170) Click **OK** ✔ from the Drawing View3 PropertyManager. View the results.

Save the Bushing part drawing.

171) Click **Save** 💾.

💡 Drawing views and notes outside the sheet boundary do not print.

Modify the Sheet Scale.

172) Right-click **Properties** in the Drawing Sheet. Expand the drop-down menu if needed. The Sheet Properties dialog box is displayed.

173) Enter Scale **3:1**.

174) Click **Apply Changes**.

Save the Bushing part drawing.

175) Click **Save** 💾. As an exercise, change the DWG. NO to 444-55 as shown below. Use the drawing Custom Properties, and the Link to Properties tool.

Summary

You learned about the Drawing and Dimension Fundamentals and created two new drawings with user defined document properties:

- Fly Wheel Assembly.

- Bushing.

You created the Fly Wheel Assembly drawing with an Exploded Isometric view.

You utilized a Bill of Materials, Magnetic lines, Balloons and learned about Custom Properties and the Title Block.

You created the Bushing Part drawing utilizing Third Angle Projection with standard Orthographic views: Front, Top, Right and Isometric.

You addressed imported dimensions from the Model Items tool and then inserted additional dimensions using the Smart Dimension tool along with all needed annotations.

A multi-view drawing should have the minimum number of views necessary to describe an object completely. The most descriptive views are those that reveal the most information about the features, with the fewest features hidden from view.

Questions

1. Describe a Bill of Materials and its contents.

2. Name the two major design modes used to develop a drawing in SOLIDWORKS.

3. True or False. Units, Dimensioning Standards, Arrow size, Font size are modified in the Options, Document Properties section.

4. How do you save a Sheet Format?

5. Identify seven components that are commonly found in a title block.

6. Describe the procedure to insert an Isometric view to the drawing.

7. In SOLIDWORKS, a drawing file name ends with a _____ suffix. A part file name ends with a _____ suffix.

8. True or False. In SOLIDWORKS, if a part is modified, the drawing is updated with a Rebuild command.

9. True or False. In SOLIDWORKS, when a dimension in the drawing is modified, the part is updated with a Rebuild command.

10. Name three guidelines to create General Notes on a drawing.

11. True or False. Most engineering drawings use the following font: Times New Roman - all small letters.

12. What are Leader lines? Provide an example.

13. Describe the key differences between a Detail view and a Section view on a drawing.

14. Identify the procedure to create an Exploded view.

15. Describe the purpose of a Design Table in a part and in an assembly.

16. Review the Design Intent section in the Introduction. Identify how you incorporated design intent into the drawing.

17. Identify how you incorporate design intent into configurations with a Design Table.

Exercises

Exercise 5.1: FLATBAR - 3 HOLE Drawing

Create the A (ANSI) Landscape - IPS - Third Angle 3HOLES drawing as illustrated below. Do not display Tangent Edges. Do not dimension to Hidden Lines.

- First create the part from the drawing, then create the drawing. Use the default A (ANSI) Landscape Sheet Format/Size.

- Insert a Front, Top and a Shaded Isometric view as illustrated. Insert dimensions. Address extension line gaps. Address proper display modes.

- Add a Smart (Linked) Parametric note for MATERIAL THICKNESS in the drawing as illustrated. Hide the dimension in the Top view. Insert Centerlines.

- Modify the Hole dimension text to include 3X THRU EQ. SP. and 2X as illustrated.

- Insert Company and Third Angle projection icons. The icons are available in the homework folder.

- Insert Custom Properties: Description, Material, DrawnBy, CompanyName, Number, Revision, and DrawnDate.

- Material is 1060 Alloy.

	Property Name	Type	Value / Text Expression	Evaluated Value
1	Number	Text	556-099	556-099
2	DrawnBy	Text	DCP	DCP
3	DrawnDate	Text	11-30-2019	11-30-2019
4	CompanyNam	Text	D&M ENGINEERING	D&M ENGINE
5	Description	Text	3HOLES	3HOLES
6	SWFormatSize	Text	8.5in*11in	8.5in*11in

Exercise 5.2: **CYLINDER Drawing**

Create the A (ANSI) Landscape - IPS - Third Angle CYLINDER drawing as illustrated below. Do not display Tangent Edges. Do not dimension to Hidden lines.

- First create the part from the drawing - then create the drawing. Use the default A (ANSI) Landscape Sheet Format/Size.

- Insert the Front, Right and Isometric view as illustrated. Insert dimensions. Think about the proper view for your dimensions. Address extension line gaps. Insert proper display modes.

- Insert Company and Third Angle projection icons. The icons are available in the homework folder.

	Property Name	Type	Value / Text Expression	Evaluated Value
1	Material	Text	"SW-Material@CylinderPart1.SLDPRT"	AISI 1020
2	Description	Text	CYLINDER	CYLINDER

- Insert Centerlines, Center Marks, and Annotations.

- Insert Custom Properties: Material, Description, DrawnBy, CompanyName, Number, Revision, and DrawnDate.

	Property Name	Type	Value / Text Expression	Evaluated Value
1	DrawnDate	Text	11-30-2019	11-30-2019
2	DrawnBy	Text	DCP	DCP
3	CompanyName	Text	D&M ENGINEERING	D&M ENGINEERING
4	Revision	Text	A	A
5	Number	Text	667-888	667-888
6	SWFormatSize	Text	8.5in*11in	8.5in*11in

- Material is AISI 1020.

Exercise 5.3: PRESSURE PLATE Drawing

Create the A (ANSI) Landscape - IPS - Third Angle
PRESSURE PLATE drawing. Do not display Tangent edges.
Do not dimension to hidden lines.

- First create the part from the drawing, then create the
  drawing. Use the default A (ANSI) Landscape Sheet
  Format/Size.

- Insert the Front and Right view as
  illustrated. Insert dimensions.
  Address all extension line gaps.
  Think about the proper view for
  your dimensions. Address
  proper display modes.

- Insert Company and Third
  Angle projection icons. The
  icons are available in the
  homework folder.

	Property Name	Type	Value / Text Expression	Evaluated Value
2	DrawnDate	Text	11-29-2019	11-29-2019
3	CompanyName	Text	D&M ENGINEERING	D&M ENGINEERING
4	Number	Text	55-568	55-568
5	Revision	Text	A	A
6	Description	Text	PRESSURE PLATE	PRESSURE PLATE
7	SWFormatSize	Text	8.5in*11in	8.5in*11in
8				

- Insert Centerlines, Center Marks
  and Annotations.

- Insert Custom Properties:
  Description, Material,
  DrawnBy, CompanyName,
  Number, Revision, and
  DrawnDate.

- Material is 1060 Alloy.

Exercise 5.3B TUBE Drawing

Create the A (ANSI) Landscape - IPS - Third Angle TUBE-2 drawing as illustrated below. Do not display Tangent edges. Phantom lines are fine.

- First create the part from the drawing, then create the drawing. Use the default A (ANSI) Landscape Sheet Format/Size.

- Insert views as illustrated. Insert dimensions. Address extension line gaps. Think about the proper view for your dimensions.

	Property Name	Type	Value / Text Expression	Evaluated Value
1	Material	Text	"SW-Material@New-Part2-Lab9.SLDPRT"	6061 Alloy
2	<Type a new property>			

- Insert Company and Third Angle projection icons. The icons are available in the homework folder.

- Insert Centerlines, Center Marks and Annotations.

- Address proper display modes.

- Insert Custom Properties: Description, Material, Revision, DrawnBy, DrawnDate, CompanyName, Number, etc.

- Material is 6061 Alloy.

	Property Name	Type	Value / Text Expression	Evaluated Value
1	CompanyName	Text	D&M ENGINEERING	D&M ENGINEERING
2	DrawnDate	Text	9-1-2019	9-1-2019
3	DrawnBy	Text	DCP	DCP
4	Revision	Text	A	A
5	Number	Text	3334-667A	3334-667A
6	SWFormatSize	Text	8.5in*11in	8.5in*11in
7	Description	Text	TUBE-2	TUBE-2

Exercise 5.4: LINKS Assembly Drawing

- Create the LINK assembly. Utilize three different FLATBAR configurations and a SHAFT-COLLAR. The parts are in the Chapter 5 homework folder.

- Create the LINK assembly drawing as illustrated. Use the default A (ANSI) Landscape Sheet Format/Size.

- Insert Company and Third Angle projection icons. The icons are available in the homework folder.

- Remove all Tangent Edges.

- Insert Custom Properties: Description, DrawnBy, CompanyName, Number, Revision, and DrawnDate.

- Insert a Bill of Materials as illustrated with Balloons.

	Property Name	Type	Value / Text Expression	Evaluated Value
1	DrawnBy	Text	⌄DCP	DCP
2	DrawnDate	Text	⌄11-30-2019	11-30-2019
3	CompanyName	Text	⌄D&M ENGINEERING	D&M ENGINE
4	Revision	Text	⌄A	A
5	PartNo	Text	⌄9998-099	9998-099
6	Description	Text	⌄LINK-2	LINK-2
7	SWFormatSize	Text	⌄8.5in*11in	8.5in*11in

Exercise 5.5: **PLATE-1 Drawing**

Create the A (ANSI) Landscape - MMGS - Third Angle
PLATE-1 drawing as illustrated below. Do not display Tangent
edges. Do not dimension to Hidden lines.

- First create the part from the drawing, then create the
  drawing. Use the default A (ANSI) Landscape Sheet
  Format/Size.

- Insert the Front and Right view as
  illustrated. Insert dimensions. Address
  extension line gaps. Think about the
  proper view for your dimensions.
  Address display modes.

- Insert Company and Third Angle
  projection icons. The icons are
  available in the homework folder.

- Insert Centerlines, Center Marks and
  Annotations.

- Insert Custom Properties:
  Description, Material, DrawnBy,
  CompanyName, Number, Revision,
  and DrawnDate.

	Property Name	Type	Value / Text Expression	Evaluated Value
1	Description	Text	PLATE-1	PLATE-1
2	DrawnBy	Text	DCP	DCP
3	DrawnDate	Text	9-1-2019	9-1-2019
4	CompanyName	Text	D&M ENGINEERING	D&M ENGINEERING
5	Revision	Text	A	A
6	Number	Text	5444-999	5444-999
7	SWFormatSize	Text	215.9mm*279.4mm	215.9mm*279.4mm

- Material is
  1060 Alloy.

Exercise 5.6: PLATE Drawing

Create the A (ANSI) Landscape - IPS - Third Angle FLAT-PLATE drawing. Do not display Tangent edges. Phantom lines are fine.

- First create the part from the drawing, then create the drawing. Use the default A (ANSI) Landscape Sheet Format/Size.

- Insert the Front, Top, Right and Isometric views as illustrated. Insert dimensions. Think about the proper view for your dimensions. Address proper display modes.

- Insert Company and Third Angle projection icons. The icons are available in the homework folder.

- Insert Centerlines, Center Marks and Annotations.

- Insert Custom Properties: Material, Description, DrawnBy, CompanyName, Number, Revision, and DrawnDate.

- Material is 1060 Alloy.

	Property Name	Type	Value / Text Expression	Evaluated Value
1	DrawnBy	Text	DCP	DCP
2	DrawnDate	Text	11-30-2019	11-30-2019
3	CompanyName	Text	D&M ENGINEERING	D&M ENGINEERING
4	Description	Text	PLATE	PLATE
5	Revision	Text	A	A
6	Number	Text	2445-990	2445-990
7	SWFormatSize	Text	8.5in*11in	8.5in*11in
8	<Type a new property>			

Exercise 5.7: LINKAGE-2 Drawing

- Create the LINKAGE-2 assembly drawing. Utilize the default A (ANSI) Landscape Sheet Format/Size. Open the LINKAGE-2 assembly from the Chapter 5 Homework\AirCylinder 2 folder.

- Insert an Isometric shaded view of the LINKAGE-2 Assembly.

- Define the PART NO. Property and the DESCRIPTION Property for the AXLE, FLATBAR-9HOLE, FLATBAR - 3HOLE and SHAFT COLLAR.

	Property Name	Type	Value / Text Expression	Evaluated Value
1	SWFormatSize	Text	215.9mm*279.4mm	215.9mm*279.4mm
2	DrawnDate	Text	11-15-2019	11-15-2019
3	DrawnBy	Text	DCP	DCP
4	CompanyName	Text	D&M ENGINEERING	D&M ENGINEERING
5	Number	Text	9999999-99	9999999-99
6	Revision	Text	A	A
7	<Type a new property>			

- Save the LINKAGE-2 assembly to update the properties. Return to the LINKAGE-2 Drawing. Insert a Bill of Materials with Auto Balloons as illustrated.

- Insert the Company and Third Angle Projection icon.

- Insert Custom Properties: Description, DrawnBy, CompanyName, Number, Revision, and DrawnDate.

Use the Pack and Go option to save an assembly or drawing with references. The Pack and Go tool saves either to a folder or creates a zip file to e-mail. View SOLIDWORKS help for additional information.

Exercise 5.8: Vertical Section View Drawing

Create a Vertical Section view drawing. Modify the drawing view scale.

Create the A (ANSI) Landscape - IPS - Third Angle drawing as illustrated below. Do not display Tangent edges.

Insert Company and Third Angle projection icons. The icons are available in the homework folder.

Insert Custom Properties: Material, Description, DrawnBy, DrawnDate, CompanyName, etc.

1. Open Section View from the Chapter 5 Homework\Section folder.

2. Create the drawing document.

3. Insert the Front view (Drawing View1).

4. Click inside the Drawing View1 view boundary. The Drawing View1 PropertyManager is displayed.

Display the origin on Sheet1.

5. Click View ➢ Hide/Show ➢ Origins from the Menu bar.

6. Click the Section View ⇥ drawing tool. The Section View PropertyManager is displayed.

7. Click the Section tab. Click the Vertical Cutting Line button.

8. Click the origin. Note: You can select the midpoint vs. the origin.

Place the Section view.

9. Click a position to the right of Drawing View1. The Section arrows point to the right. If required, click Flip direction. Check Auto hatching from the Section View box.

10. Check the Shaded With Edges option from the Display Style box. Click OK ✔ from the Section View A-A PropertyManager. Section View A-A is created and is displayed in the Drawing FeatureManager.

11. Click inside the Drawing View1 view boundary. The Drawing View1 PropertyManager is displayed.

12. Modify the Scale to 1.5:1. Click OK ✔ from the Drawing View1 PropertyManager. Both drawing views are modified (parent and child).

Exercise 5.9: Aligned Section View Drawing
Create an Aligned Section view drawing.

Create the A (ANSI) Landscape - IPS - Third Angle drawing as illustrated below.

Insert Company and Third Angle projection icons. The icons are available in the homework folder.

Insert Custom Properties: Material, Description, DrawnBy, DrawnDate, CompanyName, etc.

1. Open Aligned Section from the Chapter 5 Homework\Aligned Section folder.

2. Create the drawing document. Insert the Front (Drawing View1) and Right view.

4. Click inside the Drawing View1 view boundary. The Drawing View1 PropertyManager is displayed. Click the Section View ⇄ drawing tool. The Section View PropertyManager is displayed.

5. Click the Half Section tab.

6. Click the Leftside Down Half Section button.

7. Select the center of the drawing view as illustrated. The Section View dialog box is displayed.

8. The LENSCAP component is displayed in the Excluded components box. Check the Auto hatching box.

9. Click OK from the Section View dialog box.

Place the Section View.

10. Click a position above Drawing View1. Click OK ✔ from the Section View A-A PropertyManager.

11. Modify the scale to 1:4. Rebuild 🛢 the drawing.

12. Move the drawing views to fit the Sheet. View the drawing FeatureManager.

13. Close the drawing.

Use Section View Assist, previously called the Section View User Interface, to add offsets to existing section views. To edit existing section views with Section View Assist: Right-click the section view or its cutting line and click **Edit Cutting Line**.

Exercise 5.10: Auxiliary View Drawing

Create an Auxiliary View drawing as illustrated from an existing Front view.

Create the A (ANSI) Landscape - IPS - Third Angle drawing as illustrated.

Insert Company and Third Angle projection icons.

The icons are available in the homework folder.

Insert Custom Properties: Material, Description, DrawnBy, DrawnDate, CompanyName etc.

	Property Name	Type	Value / Text Expression	Evaluated Value
1	Material	Text	"SW-Material@Auxiliary View.SLDPRT"	1060 Alloy
2	<Type a new property>			

1. Copy the Chapter 5 Homework folder to your hard drive.

2. Open Auxiliary view from the Chapter 4 Homework\Auxiliary folder.

3. Create the drawing document.

4. Insert the Front (Drawing View1), Top, Right and Isometric view.

5. Click the angled edge in Drawing View1.

6. Click the Auxiliary View drawing tool.

7. Click a position up and to the right.

8. Drag the A-A arrow as illustrated.

9. Drag the text off the view.

10. Click OK ✔ from the Auxiliary View PropertyManager.

11. Close the drawing.

	Property Name	Type	Value / Text Expression	Evaluated Value
1	SWFormatSize	Text	215.9mm*279.4mm	215.9mm*279.4mm
2	Revision	Text	A	A
3	PartNo	Text	3345-88	3345-88
4	DrawnBy	Text	DCP	DCP
5	Description	Text	AUXILIARY	AUXILIARY
6	CompanyName	Text	D&M ENGINEERING	D&M ENGINEERING
7	DrawnDate	Text	7-1-2019	7-1-2019

Notes:

Appendix

SOLIDWORKS Keyboard Shortcuts

Below are some of the pre-defined keyboard shortcuts in SOLIDWORKS:

Action:	Key Combination:
Model Views	
Rotate the model horizontally or vertically	**Arrow** keys
Rotate the model horizontally or vertically 90 degrees	**Shift** + **Arrow** keys
Rotate the model clockwise or counterclockwise	**Alt** + left of right **Arrow** keys
Pan the model	**Ctrl** + **Arrow** keys
Magnifying glass	**g**
Zoom in	**Shift + z**
Zoom out	**z**
Zoom to fit	**f**
Previous view	**Ctrl + Shift + z**
View Orientation	
View Orientation menu	**Spacebar**
Front view	**Ctrl + 1**
Back view	**Ctrl + 2**
Left view	**Ctrl + 3**
Right view	**Ctrl + 4**
Top view	**Ctrl + 5**
Bottom view	**Ctrl + 6**
Isometric view	**Ctrl + 7**
NormalTo view	**Ctrl + 8**
Selection Filters	
Filter edges	**e**
Filter vertices	**v**
Filter faces	**x**
Toggle Selection Filter toolbar	**F5**
Toggle selection filters on/off	**F6**
File menu items	
New SOLIDWORKS document	**Ctrl + n**
Open document	**Ctrl + o**
Open From Web Folder	**Ctrl + w**
Make Drawing from Part	**Ctrl + d**
Make Assembly from Part	**Ctrl + a**
Save	**Ctrl +s**
Print	**Ctrl + p**
Additional items	
Access online help inside of PropertyManager or dialog box	**F1**
Rename an item in the FeatureManager design tree	**F2**

Action:	Key Combination:
Rebuild the model	**Ctrl + b**
Force rebuild - Rebuild the model and all its features	**Ctrl + q**
Redraw the screen	**Ctrl + r**
Cycle between open SOLIDWORKS document	**Ctrl + Tab**
Line to arc/arc to line in the Sketch	**a**
Undo	**Ctrl + z**
Redo	**Ctrl + y**
Cut	**Ctrl + x**
Copy	**Ctrl + c**
Paste	**Ctrl + v**
Delete	**Delete**
Next window	**Ctrl + F6**
Close window	**Ctrl + F4**
View previous tools	**s**
Selects all text inside an Annotations text box	**Ctrl + a**

In a sketch, the **Esc** key un-selects geometry items currently selected in the Properties box and Add Relations box.

In the model, the **Esc** key closes the PropertyManager and cancels the selections.

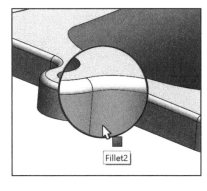

Use the **g** key to activate the Magnifying glass tool. Use the Magnifying glass tool to inspect a model and make selections without changing the overall view.

Use the **s** key to view/access previous command tools in the Graphics window.

Modeling - Best Practices

Best practices are simply ways of bringing about better results in easier, more reliable ways. The Modeling - Best Practice list is a set of rules helpful for new users and users who are trying to experiment with the limits of the software.

These rules are not inflexible, but conservative starting places; they are concepts that you can default to, but that can be broken if you have good reason. The following is a list of suggested best practices:

- Create a folder structure (parts, drawings, assemblies, simulations, etc.). Organize into project or file folders.

- Construct sound document templates. The document template provides the foundation that all models are built on. This is especially important if working with other SOLIDWORKS users on the same project; it will ensure consistency across the project.

- Generate unique part filenames. SOLIDWORKS assemblies and drawings may pick up incorrect references if you use parts with identical names.

- Apply Custom Properties. Custom Properties is a great way to enter text-based information into the SOLIDWORKS parts. Users can view this information from outside the file by using applications such as Windows Explorer, SOLIDWORKS Explorer, and Product Data Management (PDM) applications.

- Understand part orientation. When you create a new part or assembly, the three default Planes (Front, Right and Top) are aligned with specific views. The plane you select for the Base sketch determines the orientation.

- Learn to sketch using automatic relations.

- Limit your usage of the Fixed constraint.

- Add geometric relations, then dimensions in a 2D sketch. This keeps the part from having too many unnecessary dimensions. This also helps to show the design intent of the model. Dimension what geometry you intend to modify or adjust.

- Fully define all sketches in the model. However, there are times when this is not practical, generally when using the Spline tool to create a freeform shape.

- When possible, make relations to sketches or stable reference geometry; such as the Origin or standard planes, instead of edges or faces. Sketches are far more stable than faces, edges, or model vertices, which change their internal ID at the slightest change and may disappear entirely with fillets, chamfers, split lines, and so on.

- Do not dimension to edges created by fillets or other cosmetic or temporary features.

- Apply names to sketches, features, dimensions, and mates that help to make their function clear.

- When possible, use feature fillets and feature patterns rather than sketch fillets and sketch patterns.

- Apply the Shell feature before the Fillet feature, and the inside corners remain perpendicular.

- Apply cosmetic fillets and chamfers last in the modeling procedure.

- Combine fillets into as few fillet features as possible. This enables you to control fillets that need to be controlled separately; such as fillets to be removed and simplified configurations.

- Create a simplified configuration when building very complex parts or working with large assemblies.

- Use symmetry during the modeling process. Utilize feature patterns and mirroring when possible. Think End Conditions.

- Use global variables and equations to control commonly applied dimensions (design intent).

- Add comments to equations to document your design intent. Place a single quote (') at the end of the equation, then enter the comment. Anything after the single quote is ignored when the equation is evaluated.

- Avoid redundant mates. Although SOLIDWORKS allows some redundant mates (all except distance and angle), these mates take longer to solve and make the mating scheme harder to understand and diagnose if problems occur.

- Fix modeling errors in the part or assembly when they occur. Errors cause rebuild time to increase, and if you wait until additional errors exist, troubleshooting will be more difficult.

- Create a Library of Standardized notes and parts.

- Utilize the Rollback bar. Troubleshoot feature and sketch errors from the top of the design tree.

- Determine the static and dynamic behavior of mates in each sub-assembly before creating the top level assembly.

- Plan the assembly and sub-assemblies in an assembly layout diagram. Group components together to form smaller sub-assemblies.

- When you create an assembly document, the base component should be fixed, fully defined or mated to an axis about the assembly origin.

- In an assembly, group fasteners into a folder at the bottom of the FeatureManager. Suppress fasteners and their assembly patterns to save rebuild time and file size.

- When comparing mass, volume and other properties with assembly visualization, utilize similar units.

- Use limit mates sparingly because they take longer to solve and whenever possible, mate all components to one or two fixed components or references. Long chains of components take longer to solve and are more prone to mate errors.

Helpful On-line Information

The SOLIDWORKS URL:
http://www.SOLIDWORKS.com
contains information on Local
Resellers, Solution Partners,
Certifications, SOLIDWORKS users
groups and more.

Access 3D ContentCentral using the
Task Pane to obtain engineering
electronic catalog model and part
information.

Use the SOLIDWORKS Resources tab in the Task Pane to obtain access to Customer
Portals, Discussion Forums, User Groups, Manufacturers, Solution Partners, Labs and
more.

Helpful on-line SOLIDWORKS information is available from the following URLs:

*   http://www.swugn.org/

List of all SOLIDWORKS User groups.

*   https://www.solidworks.com/sw/education/certification-programs-cad-students.htm

The SOLIDWORKS Academic
Certification Programs.

*   http://www.solidworks.com/sw/indus
    tries/education/engineering-
    education-software.htm

The SOLIDWORKS Education
Program.

*   https://solidworks.virtualtester.com/#
    home_button

The SOLIDWORKS Certification Center - Virtual tester site.

*On-line tutorials are for educational purposes only. Tutorials are copyrighted by their respective owners.

SOLIDWORKS Document Types

SOLIDWORKS has three main document file types: Part, Assembly and Drawing, but there are many additional supporting types that you may want to know. Below is a brief list of these supporting file types:

Design Documents	Description
.sldprt	SOLIDWORKS Part document
.slddrw	SOLIDWORKS Drawing document
.sldasm	SOLIDWORKS Assembly document

Templates and Formats	Description
.asmdot	Assembly Template
.asmprp	Assembly Template Custom Properties tab
.drwdot	Drawing Template
.drwprp	Drawing Template Custom Properties tab
.prtdot	Part Template
.prtprp	Part Template Custom Properties tab
.sldtbt	General Table Template
.slddrt	Drawing Sheet Template
.sldbombt	Bill of Materials Template (Table-based)
.sldholtbt	Hole Table Template
.sldrevbt	Revision Table Template
.sldwldbt	Weldment Cutlist Template
.xls	Bill of Materials Template (Excel-based)

Library Files	Description
.sldlfp	Library Part file
.sldblk	Blocks

Other	Description
.sldstd	Drafting standard
.sldmat	Material Database
.sldclr	Color Palette File
.xls	Sheet metal gauge table

GLOSSARY

Alphabet of Lines: Each line on a technical drawing has a definite meaning and is drawn in a certain way. The line conventions recommended by the American National Standards Institute (ANSI) are presented in this text.

Alternate Position View: A drawing view superimposed in phantom lines on the original view. Utilized to show range of motion of an assembly.

Anchor Point: The origin of the Bill of Material in a sheet format.

Annotation: An annotation is a text note or a symbol that adds specific information and design intent to a part, assembly, or drawing. Annotations in a drawing include specific note, hole callout, surface finish symbol, datum feature symbol, datum target, geometric tolerance symbol, weld symbol, balloon and stacked balloon, center mark, centerline marks, area hatch and block.

ANSI: American National Standards Institute.

Area Hatch: Apply a crosshatch pattern or solid fill to a model face, to a closed sketch profile, or to a region bounded by a combination of model edges and sketch entities. Area hatch can be applied only in drawings.

ASME: American Society of Mechanical Engineering, publisher of ASME Y14 Engineering Drawing and Documentation Practices that controls drawing, dimensioning and tolerancing.

Assembly: An assembly is a document in which parts, features and other assemblies (sub-assemblies) are put together. A part in an assembly is called a component. Adding a component to an assembly creates a link between the assembly and the component. When SOLIDWORKS opens the assembly, it finds the component file to show it in the assembly. Changes in the component are automatically reflected in the assembly. The filename extension for a SOLIDWORKS assembly file name is *.sldasm.

Attachment Point: An attachment point is the end of a leader that attaches to an edge, vertex, or face in a drawing sheet.

AutoDimension: The Autodimension tool provides the ability to insert reference dimensions into drawing views such as baseline, chain, and ordinate dimensions.

Auxiliary View: An Auxiliary View is similar to a Projected View, but it is unfolded normal to a reference edge in an existing view.

AWS: American Welding Society, publisher of AWS A2.4, Standard Location of Elements of a Welding Symbol.

Axonometric Projection: A type of parallel projection, more specifically a type of orthographic projection, used to create a pictorial drawing of an object, where the object is rotated along one or more of its axes relative to the plane of projection.

Balloon: A balloon labels the parts in the assembly and relates them to item numbers on the bill of materials (BOM) added in the drawing. The balloon item number corresponds to the order in the Feature Tree. The order controls the initial BOM Item Number.

Baseline Dimensions: Dimensions referenced from the same edge or vertex in a drawing view.

Bill of Materials: A table inserted into a drawing to keep a record of the parts and materials used in an assembly.

Block: A symbol in the drawing that combines geometry into a single entity.

BOM: Abbreviation for Bill of Materials.

Broken-out Section: A broken-out section exposes inner details of a drawing view by removing material from a closed profile. In an assembly, the Broken-out Section displays multiple components.

CAD: The use of computer technology for the design of objects, real or virtual. CAD often involves more than just shapes.

Cartesian Coordinate System: Specifies each point uniquely in a plane by a pair of numerical coordinates, which are the signed distances from the point to two fixed perpendicular directed lines, measured in the same unit of length. Each reference line is called a coordinate axis or just axis of the system, and the point where they meet is its origin.

Cell: Area to enter a value in an EXCEL spreadsheet, identified by a Row and Column.

Center Mark: A cross that marks the center of a circle or arc.

Centerline: An axis of symmetry in a sketch or drawing displayed in a phantom font.

CommandManager: The CommandManager is a Context-sensitive toolbar that dynamically updates based on the toolbar you want to access. By default, it has toolbars embedded in it based on the document type. When you click a tab below the Command Manager, it updates to display that toolbar. For example, if you click the Sketch tab, the Sketch toolbar is displayed.

Component: A part or sub-assembly within an assembly.

ConfigurationManager: The ConfigurationManager is located on the left side of the SOLIDWORKS window and provides the means to create, select and view multiple configurations of parts and assemblies in an active document. You can split the

ConfigurationManager and either display two ConfigurationManager instances, or combine the ConfigurationManager with the FeatureManager design tree, PropertyManager or third party applications that use the panel.

Configurations: Variations of a part or assembly that control dimensions, display and state of a model.

Coordinate System: SOLIDWORKS uses a coordinate system with origins. A part document contains an original origin. Whenever you select a plane or face and open a sketch, an origin is created in alignment with the plane or face. An origin can be used as an anchor for the sketch entities, and it helps orient perspective of the axes. A three-dimensional reference triad orients you to the X, Y, and Z directions in part and assembly documents.

Copy and Paste: Utilize copy/paste to copy views from one sheet to another sheet in a drawing or between different drawings.

Cosmetic Thread: An annotation that represents threads.

Crosshatch: A pattern (or fill) applied to drawing views such as section views and broken-out sections.

Cursor Feedback: The system feedback symbol indicates what you are selecting or what the system is expecting you to select. As you move the mouse pointer across your model, system feedback is provided.

Datum Feature: An annotation that represents the primary, secondary and other reference planes of a model utilized in manufacturing.

Depth: The horizontal (front to back) distance between two features in frontal planes. Depth is often identified in the shop as the thickness of a part or feature.

Design Table: An Excel spreadsheet that is used to create multiple configurations in a part or assembly document.

Detail View: A portion of a larger view, usually at a larger scale than the original view. Create a detail view in a drawing to display a portion of a view, usually at an enlarged scale. This detail may be of an orthographic view, a non-planar (isometric) view, a section view, a crop view, an exploded assembly view or another detail view.

Detailing: Detailing refers to the SOLIDWORKS module used to insert, add and modify dimensions and notes in an engineering drawing.

Dimension Line: A line that references dimension text to extension lines indicating the feature being measured.

Dimension Tolerance: Controls the dimension tolerance values and the display of non-integer dimensions. The tolerance types are *None, Basic, Bilateral, Limit, Symmetric, MIN, MAX, Fit, Fit with tolerance* or *Fit (tolerance only).*

Dimension: A value indicating the size of the 2D sketch entity or 3D feature. Dimensions in a SOLIDWORKS drawing are associated with the model, and changes in the model are reflected in the drawing, if you DO NOT USE DimXpert.

Dimensioning Standard - Metric: - ASME standards for the use of metric dimensioning required all the dimensions to be expressed in millimeters (mm). The (mm) is not needed on each dimension, but it is used when a dimension is used in a notation. No trailing zeroes are used. The Metric or International System of Units (S.I.) unit system in drafting is also known as the Millimeter, Gram Second (MMGS) unit system.

Dimensioning Standard - U.S: - ASME standard for U.S. dimensioning uses the decimal inch value. When the decimal inch system is used, a zero is not used to the left of the decimal point for values less than one inch, and trailing zeroes are used. The U.S. unit system is also known as the Inch, Pound, Second (IPS) unit system.

DimXpert for Parts: A set of tools that applies dimensions and tolerances to parts according to the requirements of the ASME Y.14.41-2009 standard.

DimXpertManager: The DimXpertManager lists the tolerance features defined by DimXpert for a part. It also displays DimXpert tools that you use to insert dimensions and tolerances into a part. You can import these dimensions and tolerances into drawings. DimXpert is not associative.

Document: In SOLIDWORKS, each part, assembly, and drawing is referred to as a document, and each document is displayed in a separate window.

Drawing Sheet: A page in a drawing document.

Drawing Template: A document that is the foundation of a new drawing. The drawing template contains document properties and user-defined parameters such as sheet format. The extension for the drawing template filename is .DRWDOT.

Drawing: A 2D representation of a 3D part or assembly. The extension for a SOLIDWORKS drawing file name is .SLDDRW. Drawing refers to the SOLIDWORKS module used to insert, add, and modify views in an engineering drawing.

Edit Sheet Format: The drawing sheet contains two modes. Utilize the Edit Sheet Format command to add or modify notes and Title block information. Edit in the Edit Sheet Format mode.

Edit Sheet: The drawing sheet contains two modes. Utilize the Edit Sheet command to insert views and dimensions.

eDrawing: A compressed document that does not require the referenced part or assembly. eDrawings are animated to display multiple views in a drawing.

Empty View: An Empty View creates a blank view not tied to a part or assembly document.

Engineering Graphics: Translates ideas from design layouts, specifications, rough sketches, and calculations of engineers & architects into working drawings, maps, plans and illustrations which are used in making products.

Equation: Creates a mathematical relation between sketch dimensions, using dimension names as variables, or between feature parameters, such as the depth of an extruded feature or the instance count in a pattern.

Exploded view: A configuration in an assembly that displays its components separated from one another.

Export: The process to save a SOLIDWORKS document in another format for use in other CAD/CAM, rapid prototyping, web or graphics software applications.

Extension Line: The line extending from the profile line indicating the point from which a dimension is measured.

Extruded Cut Feature: Projects a sketch perpendicular to a Sketch plane to remove material from a part.

Face: A selectable area (planar or otherwise) of a model or surface with boundaries that help define the shape of the model or surface. For example, a rectangular solid has six faces.

Family Cell: A named empty cell in a Design Table that indicates the start of the evaluated parameters and configuration names. Locate Comments in a Design Table to the left or above the Family Cell.

Fasteners: Includes Bolts and nuts (threaded), Set screws (threaded), Washers, Keys, and Pins to name a few. Fasteners are not a permanent means of assembly such as welding or adhesives.

Feature: Features are geometry building blocks. Features add or remove material. Features are created from 2D or 3D sketched profiles or from edges and faces of existing geometry.

FeatureManager: The FeatureManager design tree located on the left side of the SOLIDWORKS window provides an outline view of the active part, assembly, or drawing. This makes it easy to see how the model or assembly was constructed or to examine the various sheets and views in a drawing. The FeatureManager and the Graphics window are dynamically linked. You can select features, sketches, drawing views and construction geometry in either pane.

First Angle Projection: In First Angle Projection the Top view is looking at the bottom of the part. First Angle Projection is used in Europe and most of the world. However, America and Australia use a method known as Third Angle Projection.

Fully defined: A sketch where all lines and curves in the sketch, and their positions, are described by dimensions or relations, or both, and cannot be moved. Fully defined sketch entities are shown in black.

Foreshortened radius: Helpful when the centerpoint of a radius is outside of the drawing or interferes with another drawing view: Broken Leader.

Foreshortening: The way things appear to get smaller in both height and depth as they recede into the distance.

French curve: A template made out of plastic, metal or wood composed of many different curves. It is used in manual drafting to draw smooth curves of varying radii.

Fully Defined: A sketch where all lines and curves in the sketch, and their positions, are described by dimensions or relations, or both, and cannot be moved. Fully defined sketch entities are displayed in black.

Geometric Tolerance: A set of standard symbols that specify the geometric characteristics and dimensional requirements of a feature.

Glass Box method: A traditional method of placing an object in an *imaginary glass box* to view the six principle views.

Global Coordinate System: Directional input refers by default to the Global coordinate system (X-, Y- and Z-), which is based on Plane1 with its origin located at the origin of the part or assembly.

Graphics Window: The area in the SOLIDWORKS window where the part, assembly, or drawing is displayed.

Grid: A system of fixed horizontal and vertical divisions.

Handle: An arrow, square or circle that you drag to adjust the size or position of an entity such as a view or dimension.

Heads-up View Toolbar: A transparent toolbar located at the top of the Graphic window.

Height: The vertical distance between two or more lines or surfaces (features) which are in horizontal planes.

Hidden Lines Removed (HLR): A view mode. All edges of the model that are not visible from the current view angle are removed from the display.

Hidden Lines Visible (HLV): A view mode. All edges of the model that are not visible from the current view angle are shown gray or dashed.

Hole Callouts: Hole callouts are available in drawings. If you modify a hole dimension in the model, the callout updates automatically in the drawing if you did not use DimXpert.

Hole Table: A table in a drawing document that displays the positions of selected holes from a specified origin datum. The tool labels each hole with a tag. The tag corresponds to a row in the table.

Import: The ability to open files from other software applications into a SOLIDWORKS document. The A-size sheet format was created as an AutoCAD file and imported into SOLIDWORKS.

Isometric Projection: A form of graphical projection, more specifically, a form of axonometric projection. It is a method of visually representing three-dimensional objects in two dimensions, in which the three coordinate axes appear equally foreshortened and the angles between any two of them are 120º.

Layers: Simplifies a drawing by combining dimensions, annotations, geometry and components. Properties such as display, line style and thickness are assigned to a named layer.

Leader: A solid line created from an annotation to the referenced feature.

Line Format: A series of tools that controls Line Thickness, Line Style, Color, Layer and other properties.

Local (Reference) Coordinate System: Coordinate system other than the Global coordinate system. You can specify restraints and loads in any desired direction.

Lock Sheet Focus: Adds sketch entities and annotations to the selected sheet. Double-click the sheet to activate Lock Sheet Focus. To unlock a sheet, right-click and select Unlock Sheet Focus or double click inside the sheet boundary.

Lock View Position: Secures the view at its current position in the sheet. Right-click in the drawing view to Lock View Position. To unlock a view position, right-click and select Unlock View Position.

Mass Properties: The physical properties of a model based upon geometry and material.

Menus: Menus provide access to the commands that the SOLIDWORKS software offers. Menus are Context-sensitive and can be customized through a dialog box.

Model Item: Provides the ability to insert dimensions, annotations, and reference geometry from a model document (part or assembly) into a drawing.

Model View: A specific view of a part or assembly. Standard named views are listed in the view orientation dialog box such as isometric or front. Named views can be user-defined names for a specific view.

Model: 3D solid geometry in a part or assembly document. If a part or assembly document contains multiple configurations, each configuration is a separate model.

Motion Studies: Graphical simulations of motion and visual properties with assembly models. Analogous to a configuration, they do not actually change the original assembly model or its properties. They display the model as it changes based on simulation elements you add.

Mouse Buttons: The left, middle, and right mouse buttons have distinct meanings in SOLIDWORKS. Use the middle mouse button to rotate and Zoom in/out on the part or assembly document.

Oblique Projection: A simple type of graphical projection used for producing pictorial, two-dimensional images of three-dimensional objects.

OLE (Object Linking and Embedding): A Windows file format. A company logo or EXCEL spreadsheet placed inside a SOLIDWORKS document are examples of OLE files.

Ordinate Dimensions: Chain of dimensions referenced from a zero ordinate in a drawing or sketch.

Origin: The model origin is displayed in blue and represents the (0,0,0) coordinate of the model. When a sketch is active, a sketch origin is displayed in red and represents the (0,0,0) coordinate of the sketch. Dimensions and relations can be added to the model origin but not to a sketch origin.

Orthographic Projection: A means of representing a three-dimensional object in two dimensions. It is a form of parallel projection, where the view direction is orthogonal to the projection plane, resulting in every plane of the scene appearing in affine transformation on the viewing surface.

Parametric Note: A Note annotation that links text to a feature dimension or property value.

Parent View: A Parent view is an existing view on which other views are dependent.

Part Dimension: Used in creating a part, they are sometimes called construction dimensions.

Part: A 3D object that consists of one or more features. A part inserted into an assembly is called a component. Insert part views, feature dimensions and annotations into 2D drawing. The extension for a SOLIDWORKS part filename is .SLDPRT.

Perspective Projection: The two most characteristic features of perspective are that objects are drawn: smaller as their distance from the observer increases and Foreshortened: the size of an object's dimensions along the line of sight are relatively shorter than dimensions across the line of sight.

Plane: To create a sketch, choose a plane. Planes are flat and infinite. Planes are represented on the screen with visible edges.

Precedence of Line Types: When obtaining orthographic views, it is common for one type of line to overlap another type. When this occurs, drawing conventions have established an order of precedence.

Precision: Controls the number of decimal places displayed in a dimension.

Projected View: Projected views are created for Orthogonal views using one of the following tools: Standard 3 View, Model View or the Projected View tool from the View Layout toolbar.

Properties: Variables shared between documents through linked notes.

PropertyManager: Most sketch, feature, and drawing tools in SOLIDWORKS open a PropertyManager located on the left side of the SOLIDWORKS window. The PropertyManager displays the properties of the entity or feature so you specify the properties without a dialog box covering the Graphics window.

RealView: Provides a simplified way to display models in a photo-realistic setting using a library of appearances and scenes. RealView requires graphics card support and is memory intensive.

Rebuild: A tool that updates (or regenerates) the document with any changes made since the last time the model was rebuilt. Rebuild is typically used after changing a model dimension.

Reference Dimension: Dimensions added to a drawing document are called Reference dimensions and are driven; you cannot edit the value of reference dimensions to modify the model. However, the values of reference dimensions change when the model dimensions change.

Relation: A relation is a geometric constraint between sketch entities or between a sketch entity and a plane, axis, edge or vertex.

Relative view: The Relative View defines an Orthographic view based on two orthogonal faces or places in the model.

Revision Table: The Revision Table lists the Engineering Change Orders (ECO), in a table form, issued over the life of the model and the drawing. The current Revision letter or number is placed in the Title block of the Drawing.

Right-Hand Rule: Is a common mnemonic for understanding notation conventions for vectors in 3 dimensions.

Rollback: Suppresses all items below the rollback bar.

Scale: A relative term meaning "size" in relationship to some system of measurement.

Section Line: A line or centerline sketched in a drawing view to create a section view.

Section Scope: Specifies the components to be left uncut when you create an assembly drawing section view.

Section View: You create a section view in a drawing by cutting the parent view with a cutting, or section line. The section view can be a straight cut section or an offset section defined by a stepped section line. The section line can also include concentric arcs. Create a Section View in a drawing by cutting the Parent view with a section line.

Sheet Format: A document that contains the following: page size and orientation, standard text, borders, logos, and Title block information. Customize the Sheet format to save time. The extension for the Sheet format filename is .SLDDRT.

Sheet Properties: Sheet Properties display properties of the selected sheet. Sheet Properties define the following: Name of the Sheet, Sheet Scale, Type of Projection (First angle or Third angle), Sheet Format, Sheet Size, View label, and Datum label.

Sheet: A page in a drawing document.

Silhouette Edge: A curve representing the extent of a cylindrical or curved face when viewed from the side.

Sketch: The name to describe a 2D profile is called a sketch. 2D sketches are created on flat faces and planes within the model. Typical geometry types are lines, arcs, corner rectangles, circles, polygons, and ellipses.

Spline: A sketched 2D or 3D curve defined by a set of control points.

Stacked Balloon: A group of balloons with only one leader. The balloons can be stacked vertically (up or down) or horizontally (left or right).

Standard views: The three orthographic projection views, Front, Top and Right positioned on the drawing according to First angle or Third angle projection.

Suppress: Removes an entity from the display and from any calculations in which it is involved. You can suppress features, assembly components, and so on. Suppressing an entity does not delete the entity; you can unsuppress the entity to restore it.

Surface Finish: An annotation that represents the texture of a part.

System Feedback: Feedback is provided by a symbol attached to the cursor arrow indicating your selection. As the cursor floats across the model, feedback is provided in the form of symbols riding next to the cursor.

System Options: System Options are stored in the registry of the computer. System Options are not part of the document. Changes to the System Options affect all current and future documents. There are hundreds of Systems Options.

Tangent Edge: The transition edge between rounded or filleted faces in hidden lines visible or hidden lines removed modes in drawings.

Task Pane: The Task Pane is displayed when you open the SOLIDWORKS software. It contains the following tabs: SOLIDWORKS Resources, Design Library, File Explorer, Search, View Palette, Document Recovery and RealView/PhotoWorks.

Templates: Templates are part, drawing and assembly documents that include user-defined parameters and are the basis for new documents.

Third Angle Projection: In Third angle projection the Top View is looking at the Top of the part. First Angle Projection is used in Europe and most of the world. America and Australia use the Third Angle Projection method.

Thread Class or Fit: Classes of fit are tolerance standards; they set a plus or minus figure that is applied to the pitch diameter of bolts or nuts. The classes of fit used with almost all bolts sized in inches are specified by the ANSI/ASME Unified Screw Thread standards (which differ from the previous American National standards).

Thread Lead: The distance advanced parallel to the axis when the screw is turned one revolution. For a single thread, lead is equal to the pitch; for a double thread, lead is twice the pitch.

Tolerance: The permissible range of variation in a dimension of an object. Tolerance may be specified as a factor or percentage of the nominal value, a maximum deviation from a nominal value, an explicit range of allowed values, be specified by a note or published standard with this information, or be implied by the numeric accuracy of the nominal value.

Toolbars: The toolbar menus provide shortcuts enabling you to access the most frequently used commands. Toolbars are Context-sensitive and can be customized through a dialog box.

T-Square: A technical drawing instrument, primarily a guide for drawing horizontal lines on a drafting table. It is used to guide the triangle that draws vertical lines. Its name comes from the general shape of the instrument where the horizontal member of the T slides on the side of the drafting table. Common lengths are 18", 24", 30", 36" and 42".

Under-defined: A sketch is under defined when there are not enough dimensions and relations to prevent entities from moving or changing size.

Units: Used in the measurement of physical quantities. Decimal inch dimensioning and Millimeter dimensioning are the two types of common units specified for engineering parts and drawings.

Vertex: A point at which two or more lines or edges intersect. Vertices can be selected for sketching, dimensioning, and many other operations.

View Palette: Use the View Palette, located in the Task Pane, to insert drawing views. It contains images of standard views, annotation views, section views, and flat patterns (sheet metal parts) of the selected model. You can drag views onto the drawing sheet to create a drawing view.

Weld Bead: An assembly feature that represents a weld between multiple parts.

Weld Finish: A weld symbol representing the parameters you specify.

Weld Symbol: An annotation in the part or drawing that represents the parameters of the weld.

Width: The horizontal distance between surfaces in profile planes. In the machine shop, the terms length and width are used interchangeably.

Zebra Stripes: Simulate the reflection of long strips of light on a very shiny surface. They allow you to see small changes in a surface that may be hard to see with a standard display.

INDEX

About Us

SDC Publications specializes in creating exceptional books that are designed to seamlessly integrate into courses or help the self learner master new skills. Our commitment to meeting our customer's needs and keeping our books priced affordably are just some of the reasons our books are being used by nearly 1,200 colleges and universities across the United States and Canada.

SDC Publications is a family owned and operated company that has been creating quality books since 1985. All of our books are proudly printed in the United States.

Our technology books are updated for every new software release so you are always up to date with the newest technology. Many of our books come with video enhancements to aid students and instructor resources to aid instructors.

Take a look at all the books we have to offer you by visiting SDCpublications.com.

NEVER STOP LEARNING

Keep Going

Take the skills you learned in this book to the next level or learn something brand new. SDC Publications offers books covering a wide range of topics designed for users of all levels and experience. As you continue to improve your skills, SDC Publications will be there to provide you the tools you need to keep learning. Visit SDCpublications.com to see all our most current books.

Why SDC Publications?

- Regular and timely updates
- Priced affordably
- Designed for users of all levels
- Written by professionals and educators
- We offer a variety of learning approaches

TOPICS
3D Animation
BIM
CAD
CAM
Engineering
Engineering Graphics
FEA / CAE
Interior Design
Programming

SOFTWARE
Adams
ANSYS
AutoCAD
AutoCAD Architecture
AutoCAD Civil 3D
Autodesk 3ds Max
Autodesk Inventor
Autodesk Maya
Autodesk Revit
CATIA
Creo Parametric
Creo Simulate
Draftsight
LabVIEW
MATLAB
NX
OnShape
SketchUp
SOLIDWORKS
SOLIDWORKS Simulation